An Account Of The Battle Of Wilson's Creek

By

Thomas W. Adams

R. I. Holcombe

An Account Of The Battle Of Wilson's Creek

by Thomas W. Adams
R. I. Holcombe

ISBN: 978-93-59950-53-2
Published by

DOUBLE 9 BOOKS

2/13-B, Ansari Road
Daryaganj, New Delhi – 110002
info@double9books.com
www.double9books.com
Tel. 011-40042856

ABOUT THE AUTHOR

Thomas W. Adams, the author of "AN ACCOUNT OF THE Battle of Wilson's Creek," is well known for this specific paintings as considered one of his masterpieces. As a wonderful author, Adams masterfully bridges the geographical regions of military and history, weaving problematic narratives that captivate and teach his readers. His words function a conduit, forging connections amongst humans with the aid of fostering a deeper expertise of the beyond, particularly within the context of navy activities just like the Battle of Wilson's Creek. Adams' writing is a testament to his boundless creativity and unwavering passion for his craft. He expertly courses readers thru a diverse range of problem depend and emotions, making his works a wealthy tapestry of reports. In his prose, beauty and accessibility coexist harmoniously, inviting a wide target audience to revel in the high-quality tales he stocks. Whether delving into navy records or other themes, Thomas W. Adams' writings provide a gateway to exploration and enlightenment. His potential to combine the beauty of storytelling with the simplicity of comprehension guarantees that readers from all walks of existence can partake in the magic of his narratives. In essence, Adams' literary contributions are a bridge to the beyond and a supply of leisure for all who immerse themselves in his paintings.

CONTENTS

CHAPTER I
PRELIMINARY MOVEMENTS TO THE BATTLE

In giving an account of the battle of Wilson's Creek, or Oak Hills, which though not the largest, has passed into history, as one of the hardest and best fought battles of the American Civil War, it is necessary to describe certain military movements and operations which took place previously, in order that a better understanding of all of the circumstances may be had. This must be done here briefly and in a somewhat desultory way.

Upon the outbreak of the civil war in 1861, the people of Southwest Missouri were divided in sentiment, although a majority of them were Unionists. At the previous Presidential election, Lincoln, the Republican candidate, had received 42 votes in Greene county alone, and this district had sent unconditional Union candidates to the State Convention the previous February by a vote of four to one. Union Home Guards were organized in Springfield in May to the number of 1200, composed of citizens of Greene, Christian, and adjoining counties and commanded by Col. John S. Phelps (afterward Governor). The secessionists in this quarter of the State were in the minority, but they were bold and disposed to be aggressive.

FIRST FEDERAL TROOPS.

In a few days after the occupation of Rolla, Col. Franz Sigel took up the line of march for Springfield. He had his own regiment, the 3d Mo. Volunteers, and Col. Chas. E. Salomon's 5th Missouri Volunteers. The march from Rolla to Springfield was necessarily slow, as the Federals were compelled to feel their way cautiously, but, considering all of the circumstances, very good time was made. Detachments were sent out on either side of the road from time to time, and the country pretty well reconnoitered.

"SIGEL IS COMING!"

At last, on Sunday morning, June 24, 1861, the citizens of Springfield who lived in the eastern part of town, looked out on the St. Louis road and saw, coming leisurely along, a column of men led by others on horseback. The wind lifted and shook out a banner, which, when unfolded, showed the old familiar stripes in all their splendor and the stars in all their beauty. Just then the band struck up a spirit-stirring air, and the cry rang out and was caught up and borne through all the town, "They are coming! They are coming!" If it was asked, "Who are coming?" the reply sometimes was, "The Union soldiers," but often came the answer, "The Yankee Dutch." People had different ways of looking at the thing and different ideas altogether about the matter!

But whether they were "brave Union Germans" or "Yankee Dutch," certain it was that Sigel and his troops were in full possession of the town. It was about 11:30 in the forenoon when the soldiers reached the main part of town. Pickets were put out on all roads, and many prisoners made among the citizens accused of real or premeditated "treason" against the government. The court-house was pretty well filled at one time with these prisoners. Some property was seized or "pressed" by the soldiers, and their presence did not give universal or even general satisfaction.

GEN. SWEENEY COMES TO SPRINGFIELD.

On the 1st of July Gen. T. W. Sweeney (then really only a captain in the regular army), having been *elected* a brigadier by the St. Louis Home Guards, came to Springfield with a force of, say 1,500 men, including the 1st Iowa Infantry (dressed in gray uniforms) a portion of the 2d Kansas, and some artillery and a battalion of regular dragoons.

By reason of his rank, which was recognized as that of brigadier, Gen. Sweeney became the commander of the Federal army, then in Southwest Missouri. He was a brigadier-general of Home Guards or U. S. Reserve Corps; Sigel and Salomon and Brown were but colonels of volunteers. Sweeney was an Irishman. He had but one arm, having lost the other in the Mexican war. Like many another of his countrymen, he had more fight in him than good judgment. Although starting in rank pretty well at the top at the beginning of the war, he never attained any great military distinction. After the war he led the Fenian raid into Canada, which ended so ignominiously.

SIGEL DEPARTS FOR CARTHAGE.

After the battle of Boonville, June 17, the State forces, under Col. Marmaduke and Gov. Jackson, retreated toward the Southwest portion of the State to co-operate with the troops under Gen. Rains, and to be in easy distance of the Confederate forces at Fayetteville, Ark., under Gen. Ben McCulloch. News of this movement having reached Gen. Sigel at Springfield, that officer at once set out to intercept it—to prevent, if possible, a junction between the forces of Col. Marmaduke and those of Gen. Rains, and to attack the latter and destroy him in his camp, supposed to be near Rupe's Point, in Jasper county.

"Pressing" a number of horses and wagons from the citizens of this county, especially from about Springfield, Sigel, with the greater part of his own and Salomon's regiment and a company of regulars, set out from Springfield westward on the Mt. Vernon road, one hot morning about the 1st of July. His destination was Carthage, 65 miles away. He had with him eight pieces of Backof's artillery, 6 and 12 pounders. On the 5th the battle of Carthage was fought between the eight companies of Sigel's regiment, seven companies of Salomon's and the artillery under Backof on the Union side, and the State Guards under Gov. Jackson in person, and Gens. Rains and Parsons. The Federals were defeated and fell back to Mt. Vernon, Sigel being foiled in his attempt to prevent the concentration of the secessionists.

GEN. LYON ENTERS THE COUNTY.

On the 3d of July Gen. Nathaniel Lyon, at the head of about 2,000 troops, left Boonville for the Southwest to co-operate with Sigel. On the 25th of June five companies of cavalry, six companies of regular infantry and dragoons, and ten companies of Kansas volunteers, in all about 1,600 men, under command of Maj. S. D. Sturgis, left Kansas City, destined also for Southwest Missouri. At Grand River, in Henry county, the two commands formed a junction, and then started for Sigel. Hearing of the latter's defeat, and retreat to the eastward, Gen. Lyon changed his direction more to the eastward and came into this county about the 13th of July, going into camp near Pond Spring, on section 31, township 29, range 23, in the western part of the county. Lyon came into the town of Springfield July 13th, leaving, as he wrote to Chester Harding, his troops, "a few miles back."

Gen. Lyon was mounted on an iron-gray horse, and had an escort or body-guard of ten men of the 1st regiment U. S. regular cavalry, all of whom were men remarkable for their large size, strong physique, and fine horsemanship. Lyon treated the citizens with courtesy and kindness, although impressing their provisions and animals, to some extent, for the use of his men. As soon as he arrived in this quarter he communicated with Sigel, and with Gen. Fremont at St. Louis, asking the latter to send him reinforcements at once. He also busied himself in recruiting for the Federal service—issuing commissions to officers of Home Guard companies, and mustering in enlisted men. Ho was visited by Union men from counties north and east 75 miles away.

SWEENEY'S EXPEDITION TO FORSYTH.

Saturday, July 20, about 1,200 men were detailed under Gen. Sweeney to break up a secession camp reported to be at Forsyth, the county seat of

Taney county. The command was composed of the two companies of regular cavalry, under Capt. D. S. Stanley; a section of Capt. Totten's battery, in charge of Lt. Sokalski; about 500 men of the 1st Iowa Infantry, under Lt. Col. Merritt; Capt. Wood's company of mounted Kansas volunteers, and the 2d Kansas Infantry, under Col. Mitchell. The expedition reached Forsyth in the afternoon of Monday, captured the town with but little difficulty, driving out about 200 State Guards, who had been quartered in the court-house, and secured some blankets, clothing, guns, provisions, horses and one or two prisoners. A quantity of lead was taken from a well into which it had been thrown. Three shells were thrown into the court-house after the Federals had possession of the town.

Gen. Sweeney remained in Forsyth about 24 hours, and returned to Springfield on Thursday. His loss was three men wounded, and Capt. Stanley had a horse shot under him. It was reported that the secessionists had five killed and ten wounded, among the latter being one Capt. Jackson. A camp of 1,000 Confederates, at Yellville, Ark., was not molested by Gen. Sweeney, although only 50 miles from Forsyth.

CONFEDERATE MILITARY OPERATIONS.

Meantime preparations were making among the secessionists of Missouri to dispute the occupancy of the Southwest portion of the State with the Federals. Gen. Ben McCulloch, of Texas, had been ordered by the Confederate government to go to the assistance of its allies in Missouri. Accordingly he rendezvoused at Fayetteville, Ark., where he was joined by some Louisiana and Arkansas volunteers and a division of Arkansas State troops. The Missouri State Guards, Gov. Jackson's troops, had rendezvoused, first near Sarcoxie, in Jasper county, afterward on the Cowskin Prairie, in McDonald county, where some time was spent in drilling, organizing and recruiting.

On the 25th of July, 1861, General Sterling Price, in command of Gov. Jackson's State Guard, began to move his command from its encampment on the Cowskin Prairie toward Cassville, Barry county, at which place it had been agreed between Generals McCulloch and N. B. Pearce, of the Confederate force, and Price that their respective commands, together with General J. H. McBride's division of State Guards, should concentrate, preparatory to a forward movement on Lyon and Sigel and the other Federal troops in the vicinity of Springfield. On the 29th the junction was effected. The combined armies were then put under marching orders. The 1st division, commanded by Gen. McCulloch in person; the 2d, by Gen. Pearce, of Arkansas, and the 3d, by Gen. Steen, of Missouri, left Cassville on the 1st and 2d of August, taking the Springfield road. It is said that Gen. Price, with the greater portion of his infantry, accompanied the 2d division. A few days afterward a regiment of Texas rangers, under Col. E. Greer, joined the martial host advancing to attack the Federals. Gen. James S. Rains, formerly the well known politician of Jasper county, with six companies of mounted Missourians belonging to his division, the 8th, commanded the advance guard. Rains was given the advance because many of his men were from this quarter of the State and knew the country very well. On Friday, August 2, he encamped at Dug Springs, in Stone county, about 20 miles southwest of Springfield. The main army was some distance to the westward.

The Southern army was really composed of three small armies, as follows: The Missouri State Guard, under Gen. Price; a division of Arkansas State troops, under Gen. N. Bart. Pearce, and a division of Confederate troops under Gen. McCulloch. Pearce's division was composed of the 1st Arkansas cavalry, Col. De Rosey Carroll; Capt. Chas. A. Carroll's independent company of cavalry; the 3d Arkansas infantry, Col. John R. Gratiot; the 4th Arkansas infantry, Col. J. D. Walker; the 5th Arkansas infantry, Col. Tom P. Dockery, and Capt. Woodruff's battery, the "Pulaski Artillery." All of the infantry regiments had enlisted only for three months, and their time expired about Sept. 1. They were *State* troops, or militia. Another Arkansas battery, Capt. J. G. Reid's, of Ft. Smith, was also with Gen. Pearce, but assigned to McCulloch afterwards.

THE FIGHT AT DUG SPRINGS.

Gen. Lyon was duly informed of the concentration of the Southern troops at Cassville, of the junction of Price and McCulloch, and of their intention of marching upon his own camp. His scouts and spies were numerous, sharp and faithful. They marched in the ranks with the secession troops at times, hung about officers' quarters, picked up all the information they could and then made their way inside the Federal lines in a very short time. For the most part Lyon's scouts were residents of this part of the State and knew all the country very thoroughly. Gen. Price, too, had scouts and spies, who kept *him* posted—who, by various ruses and stratagems visited the Federal camps, and obtained valuable information and conveyed it to "Old Pap" in short order. And Price's scouts, too, were chiefly residents of Southwest Missouri. A number of Greene county men did scouting for both Price and Lyon.

Learning of the movements of Price and McCulloch, Gen. Lyon determined to go out and meet them. He first sent more messengers to Gen. Fremont, at St. Louis, begging for reinforcements, and late in the afternoon of Thursday, the 1st of August, his entire army, which, by the addition of Sigel's and Sturgis' forces, had been increased to 5,868 men of all arms, infantry, cavalry and eighteen pieces of artillery, led by himself, moved toward Cassville, leaving behind a force of volunteers and Home Guards to guard Springfield. That night the army bivouacked about ten miles southwest of Springfield, on a branch of the James. Gen. Lyon's subordinate commanders were Brig. Gen. T. W. Sweeney, Col. Sigel and Maj. Sturgis. The next morning, early, the command moved forward. It was a hot day and the men suffered severely from dust, heat and excessive thirst, most of the wells and the streams being dry. Towards evening five dollars was offered for a canteen of warm ditch water.

At Dug Springs the army halted, having come up with Gen. Rains' advance of the Southern forces. The Missourians were first observed about 11 o'clock in the forenoon, at a house by the roadside with a wagon partially laden with cooked provisions, from which they were driven away by shell from one of Capt. Totten's guns. At the Dug Springs (which are in an oblong valley, five miles in length and broken by projecting spurs of the hills, which form wooded ridges), at about 5 o'clock in the evening a skirmish took place between Rains' secessionists and a battalion of regular infantry under Capt. Fred Steele, a company of U. S. dragoons under Capt. D. S. Stanley, and two 6-pounders of Capt. Totten's battery. The Southerners were driven away with a loss of one killed, perhaps half a dozen wounded, and ten prisoners. A Lieutenant Northent is reported as having been mortally wounded. The Federal loss was four killed outright, one mortally wounded, and about thirty slightly wounded. Three of the Federal killed were Corporal Klein, and Privates Givens and Devlin.

On the side of the Missourians a young man named H. D. Fulbright, was sunstruck in the engagement, and died. W. J. Frazier, of the Greene County Company, attached to McBride's division, was wounded.

The Federals pursued next morning, going as far as Curran, or McCullah's store, nearly on the county line between Stone and Barry counties, and twenty-six miles from Springfield. During the day a scouting party of secessionists, which had come across the country from Marionville, was encountered at dinner. Totten's artillery was brought up, a few shells fired, and the Southern troops did not wait for the dessert! This is a brief, but correct account of what is often referred to in histories of the civil war as the *"battle"* of Dug Springs.

GEN. LYON FALLS BACK.

Finding that the enemy in his front was much his superior in numbers, Gen. Lyon determined to go no farther than Curran, but to return to

Springfield and await the reinforcements so urgently requested of Gen. Fremont before risking a decisive battle, the result of which would certainly mean a splendid victory and possession of all Southwestern Missouri to one party or the other. The Federal scouts also reported that a large force of State Guards was marching to the assistance of Gen. Price from toward Sarcoxie. Accordingly, after a conference with his officers, Sweeney, Sigel, and Majors Sturgis, Schofield, Shepherd, and Conant, and the artillery captains, Totten and Schaeffer, Gen. Lyon countermarched his army and returned to Springfield, coming this time directly to the town, where he arrived August 5th. The main body of the army camped about the town. Nearly 2,000 of the volunteers and regulars under Lt. Col. Andrews, of the 1st Missouri, and Maj. Sturgis were stationed out about four miles from town. Two days later this force was withdrawn to the line of defence around the town.

A vigilant guard was at once set upon all roads and avenues of approach to Springfield. No one was allowed to *go out*, except physicians, although everybody was admitted. Never, perhaps, in the history of war was a camp so well guarded, and all knowledge of its character kept so well from the enemy as was Gen. Lyon's at Springfield.

Col. Thos. L. Snead, now of New York City, and Gen. Price's assistant adjutant general in 1861, has kindly furnished much very valuable information to the writer hereof, and through this volume to the world at large. The colonel's means of knowledge are very superior, and he has manifested the utmost willingness to impart what he knows concerning the memorable days of July and August, 1861.

Col. Snead says that on Sunday morning, August 4th (1861), Gen. Price and he rode over to Gen. McCulloch's headquarters, at McCulloch's farm, and in the presence of Snead and Col. James McIntosh, who was McCulloch's adjutant general, Gen. Price urged McCulloch to co-operate with him in an attack on Lyon, who was supposed to be in the immediate front,—it not then being known to the Confederates that he had retreated. McCulloch had no faith in Price's skill as an officer, and a profound contempt for the Missouri officers generally,—and for Gen. Rains particularly.[1]

Gen. Price was a major-general of Missouri militia, McCulloch only a Confederate brigadier. Price had a loud voice and a positive address, and always spoke to McCulloch as if the latter were his inferior. "Do you mean to march on and attack Lyon, Gen. McCulloch?" he demanded. "I have not received orders yet to do so, sir," answered McCulloch; "my instructions leave me in doubt whether I will be justified in doing so." "Now, sir," said Price, still in his loud, imperious tone, "I have commanded in more battles than you ever saw, Gen. McCulloch. I have three times as many troops as you.

I am of higher rank than you are, and I am twenty years your senior in age and general experience. I waive all these considerations, Gen. McCulloch, and if you will march upon the enemy I will obey your orders, and give you the whole command and all the glory to be won there!" McCulloch said he was then expecting a dispatch from President Davis, and would take Gen. Price at his word if it should be favorable, and if after consultation with Gen. Pearce the latter should agree also to co-operate, Gen. Pearce having an independent command of Arkansas State troops.

Gen. Price immediately called his general officers together and told them what he had done. They were at first violently opposed to his action, but finally they gave their unwilling consent to what they considered an unnecessary self-abasement. In the afternoon McCulloch and McIntosh came to Price's headquarters, and McCulloch announced that he had received from Richmond, since morning, dispatches that gave him greater freedom of action and also that he would receive that night 1,000 reinforcements (Greer's Texas regiment), and that he would therefore accede to Gen. Price's proposition and assume command of the combined armies and march against Gen. Lyon. Accordingly Col. Snead wrote, by Gen. Price's direction, the necessary orders and had them published to the Missouri State Guard. It having been learned that the Federals were retreating, orders were given to move that very night. Lyon had, however, escaped, "and," says Col. Snead, "this was fortunate for us, perhaps."

THE SOUTHERN FORCES UNDER PRICE AND M'CULLOCH ENTER GREENE COUNTY—A GREAT BATTLE IMMINENT.

When Gen. Rains' troops were driven from the field at Dug Springs, they fell back on the main army under Price and McCulloch, some five miles away, and reported that the force which had assailed them was not only vastly superior to their own, but was much larger and more formidable than the combined Southern army. It was evident that Gen. Rains, if not badly whipped, was badly frightened. The Confederates and Missourians were then encamped on Crane Creek, in the northern part of Stone county.

Thereupon there was confusion among the principal Southern officers. General McCulloch counselled a retreat and General Price advocated a forward movement. Price's officers and men agreed with him and were "eager for the fray." As McCulloch was unwilling to advance, General Price asked him to loan him some arms for the destitute portion of his command, that the Missourians might advance by themselves. McCulloch at first refused; afterwards 800 muskets were furnished the Missourians. The embarrassing disagreement continued till in the evening of Sunday, August 4, when an order was received by McCulloch from the Confederate authorities ordering what Price much desired—an advance on General Lyon. A council was at once held, at which McCulloch agreed to march on Springfield provided he was granted the chief command of the consolidated army. Price, to whom in right and justice the supreme command belonged, anxious to encounter the Federals and defeat and drive them from the State before they could be reinforced by Fremont from St. Louis, consented to the terms of the imperious Texas ranger, saying: "I am not fighting for distinction, but for the liberties of my country, and I am willing to surrender not only my command but my life, if necessary, as a sacrifice to the cause." A little after midnight, therefore, on Sunday, August 4, the Southern camp was broken up and the troops took up the line of march, which was continued slowly and cautiously, along the Fayetteville road to the crossing of Wilson's Creek, near the Christian county line, in sections 25 and 26, tp. 28, range 23, ten miles southwest of Springfield, which locality was reached on the 6th.

CHAPTER II
THE BATTLE OF WILSON'S CREEK—
THE UNION OR FEDERAL ACCOUNT

GEN. LYON IN SPRINGFIELD.

When Gen. Lyon returned to Springfield after the Dug Springs expedition, he scattered his forces upon the different roads leading into the city at a distance of from three to five miles. Five miles from town, on the Fayetteville road, was a force of 2,500 under command of Maj. Sturgis. The other roads were well guarded, and all precautions were taken against a surprise or a sudden attack. Gen. Lyon's private room and personal headquarters were in a house on North Jefferson street, not far from the public square. The building, a small one, was then owned by Mrs. Boren; it is now (1883) the property of Mrs. Timmons. His general headquarters were on the north side of College street, a little west of Main, in a house then owned by John S. Phelps, but which had been recently occupied by Maj. Dorn. In this same house his body lay, after it was borne from the battle field of Wilson's Creek. The house was burned by Curtis' Federals in February, 1862, and where it once stood is now (July, 1883) a vacant lot, on which are the remains of an old cellar.

As soon as Lyon reached Springfield he again sent off a courier to Fremont at St. Louis asking for reinforcements. Hon. John S. Phelps, who had started for Washington City to attend the extra session of Congress convened by President Lincoln, had stopped in St. Louis, called upon Gen. Fremont, and urged him to help Lyon and the Union people of Southwest Missouri with men and supplies, both of which were at St. Louis in abundance.[2] But Fremont stated that he did not believe Gen. Lyon was in anything like desperate straits; that McCulloch and Price could have nothing but an inconsiderable force, since the country in Southwestern Missouri was too poor to support a force of any formidable strength; that in his opinion Lyon could take care of himself; and finally that he had no troops to spare him anyhow, as he had received information through Gov. Morton, of Indiana, that a large Confederate force and flotilla of gunboats, under command of Gen. Pillow, were coming up the Mississippi to attack

Cairo, Bird's Point, and if successful in their destruction, would come on and destroy St. Louis, and that he had need of every available man to guard those threatened points.

Gen. Lyon consulted with his officers and with the prominent Union men of Springfield very freely. He knew the situation perfectly. His scouts came in every day from McCulloch's army and gave him all needed information. He was impatient to fight the force in his front, but he anxiously desired reinforcements to enable him to have a reasonable chance of success. Every day he visited the out-posts and nearly every day sent off messages for help. Sometimes he would lose his temper and curse and swear quite violently. On one occasion he received a message from Fremont that no more troops could or would be sent for the present. Striding back and forth in his room, with the paper in his hand, he suddenly threw it on the table, and smiting his hands together cried out: "*G—d d—n General Fremont*: He is a worse enemy to me and the Union cause than Price and McCulloch and the whole d—d tribe of rebels in this part of the State!"[3]

PRELIMINARIES OF THE FINAL STRUGGLE.

On Monday, August 5, the day of Lyon's arrival at Springfield, as before stated, he left a force of 2,500 strong at a point about five miles from Springfield, on the Fayetteville road. This force (comprising fully one-third of Lyon's army), under Major Sturgis, was ordered by Gen. Lyon to be ready to move at a moment's notice, and at about 6 o'clock on the evening of the next day the men were in ranks, the artillery horses harnessed, and everything in readiness to march back and attack the advancing enemy.

Shortly afterward a stream of visitors, messengers, and communications poured in upon the general, some reporting the engagement of Capt. Stockton, of the 1st Kansas, and two companies of Home Guards with a party of Price's cavalry, on the prairie west of town, in which two of the latter were wounded; some gave other information; some were the bearers of excellent advice(!); others came for orders; still others had no business.

Two companies were ordered to the relief of Capt. Stockton. Eight companies of the 1st Kansas infantry, a part of the second Kansas, and Major Osterhaus' battalion of the 2d Missouri were ordered to a certain point in town to await the arrival of Gen. Lyon, who, it seems, was so entirely occupied with other matters that instead of starting at 10 o'clock, it was midnight when he left his headquarters, and without looking at his watch he proceeded to Camp Hunter, having already ordered Major Sturgis to drive in the enemy's pickets, if within two miles of his own. A company of cavalry under Capt. Fred Steele[4] was dispatched on his errand (to find the pickets) at half-past 12, and Gen. Lyon, with the troops above mentioned, arrived at 3 o'clock in the morning. Here he consulted his watch, and, finding the time more than two hours later than he supposed, he at once called together his principal officers, communicated to them his embarrassing position, and taking their advice, withdrew the entire force to Springfield.

It had been Lyon's intention, on retreating from Dug Springs to Springfield, to wheel suddenly about on reaching the latter place and march back upon Price and McCulloch (who, he considered, would be following him up), fall upon them when they least expected an attack, and defeat them if possible. On arriving at Springfield, appearances indicated the approach of a Confederate force from the west, and this caused him to wait a few hours. The night of the 6th, his information was to the effect that Price and McCulloch were only seven miles away from Sturgis' camp, and he intended attacking them at daylight. On the return to town the general remarked to Major Schofield, of the 1st Missouri (Frank Blair's regiment), that he had a premonition that a night attack would prove disastrous, and yet he had felt impelled to try it once, and perhaps should do so again, "for my only hope of success is in a surprise," he added. Before the Federals reached Springfield it was daylight. An ambush was formed in the timber southwest of town in case of pursuit.

During Wednesday continued alarms were circulated in Springfield, and a real panic prevailed among many of the citizens, who packed up and left, or prepared to leave, for supposed places of safety. The troops wore under arms in every quarter, and several times it was reported that fighting had actually commenced. Toward night the panic in a degree subsided: but many of the people who remained did not retire or make any attempt to sleep. Phelps' regiment of Home Guards, commanded by Col. Marcus Boyd, was on the *qui vive* the whole night.

A consultation of the principal Federal officers was held at Gen. Lyon's headquarters, which lasted till midnight. The question of evacuating Springfield and abandoning Southwest Missouri to its fate was seriously discussed. Looking at the matter from a military point of view, there was no doubt of the propriety and even the necessity of such a step, and Gen. Lyon and the majority of his officers counseled such a movement. Some favored a retreat to Fort Scott, while others thought Rolla a point easier reached and promising better results.

Gen. Sweeney, however, was strongly opposed to retreating without a fight. With his naturally florid face flushed to livid red, and waving his one arm with excitement, he exclaimed vehemently against such a policy — pointing out the disastrous results which must ensue upon a retreat without a battle—how the "rebels" would boast over such an easy conquest, how they would terrorize, harass, and persecute the unprotected Unionists if given undisputed possession of the country, how the Unionists themselves would become discouraged, crushed, or estranged, and declared himself in favor of holding on to the last moment, and of giving battle to Price and McCulloch as soon as they should offer it.[5]

Gen. Lyon and some of the other officers became converts to Gen. Sweeney's views, and it was decided to remain, save the reputation of the little army, hope against hope for reinforcements, and not evacuate Springfield and Greene county until compelled to. The next day when Sigel's brigade quartermaster, Major Alexis Mudd, asked Gen. Lyon when the army would leave Springfield, the latter replied: "Not until we are whipped out."

A FALSE ALARM.

Thursday morning, Price and McCulloch were reported to be actually advancing on Springfield. Lyon's troops were quickly in line of battle, the luggage wagons were all sent to the center of the town, and in this position they remained during nearly the entire day. The Southern troops *had* advanced, but only about two miles, and had gone into camp in the southern part of this county, nearly on the line between Greene and Christian counties (in sections 25 and 36, tp. 28, range 23, partly in Greene and partly in Christian county), their tents being on either side of Wilson's Creek, and extending a mile or so east and south of the Fayetteville road. Thursday evening the Federals were ready for marching orders, but a portion of the Kansas troops had been so much engaged the night before as to be really unfit for service, and an order for all of the soldiers, except those actually on guard, to retire and rest, was issued, and the night attack was again deferred. The Home Guards were on duty and in active service in the city at this time.

And so the soldiers lay down to rest and to sleep—to many of them it was to be the last repose they should take until they should lie down to take their last sleep. Soon the camps were wrapped in silence and slumber and no sound was to be heard save the cry of the night birds and the challenges of the watchful pickets as they hailed the relief guard, or arrested the steps of some belated wanderer. There they lay, these men from Iowa and Kansas, dreaming of the homes and loved ones they had left behind them on the beautiful prairies of their own States, and in vision seeing faces and forms and scenes they were destined to never see again in reality. There they lay, these bearded Germans from St. Louis, dreaming, perhaps, of families and kinsmen in the city by the great river, or of their early homes in the

Fatherland, far across the deep, blue sea. There they lay, these Missouri Unionists, sleeping as peacefully as their brethren in arms.

There they lay, too, only a few miles away, those men under the folds of the new flag, who had come out from their homes by the bayous of Louisiana, on the plains of Texas, amid the hills and dales and valleys of Arkansas and Missouri, to do battle for the cause they believed to be just and righteous, to drive out those whom they believed to be the wrongful invaders of their country, the despoilers of their homes. And to blue and gray alike, with an equal peace and softness, came that balmy blessing which "knits up the raveled sleeve of care."

Friday, the 9th, Springfield was remarkably quiet. But the calm preceded the storm. Those timid creatures who had made it a business to repeat exciting rumors had been frightened away with much of the material upon which they operated. Enlistments in the Springfield regiment had been rapid, and really among the uninitiated and uninformed a feeling of security prevailed. During the afternoon, Capt. Wood's company of Kansas cavalry and Capt. Stanley's company of regulars had a skirmish with a scouting party of Price's cavalry on the prairie about five miles west of town, defeating them, wounding two and capturing six or eight prisoners. From the prisoners, among other information, it was learned that the Southern troops were badly off for provisions and were forced to do some pretty liberal foraging on both friends and enemies.

A MESSENGER FROM FREMONT—NO HOPE!

About noon there arrived a messenger from St. Louis and Fremont bearing a dispatch from the latter to Gen. Lyon. This dispatch informed Lyon that his situation was not considered critical; that he doubtless over-estimated the force in his front; that he ought not to fall back without good cause, and assured him that no reinforcement would be sent, but that he must report his future movements as soon as possible, and do the best he could.

Like the brave, disciplined soldier that he was, Lyon accepted the situation, and prepared to obey the orders of his superior officer. With Fremont's message before him, he sat quietly down at his little table in his headquarters and wrote the following reply with his own hand—the last letter he ever wrote:—

Springfield, Mo., Aug. 9, 1861.

General—I have just received your note of the 6th inst., by special messenger. I retired to this place, as I have before informed you, reaching here on the 5th. The enemy followed to within ten miles of here. He has taken a strong position, and is recruiting his supplies of horses, mules, and provisions by forages into the surrounding country. His large force of mounted men enables him to do this without much annoyance from me. I find my position extremely embarrassing, and am at present unable to determine whether I shall be able to maintain my ground or forced to retire. I shall hold my ground as long as possible, though I may, without knowing how far, endanger the safety of my entire force with its valuable material, being induced, by the important considerations involved, to take this step. The enemy yesterday made a show of force about five miles distant, and has doubtless a full purpose of making an attack on me. Very respectfully, your obedient servant,

N. Lyon
Brig. Gen. Vols., Commanding.

To Major Gen. J. C. Fremont, Commanding Western Department, St. Louis, Mo.

No word of complaint; no murmuring; but with the expressed knowledge that he was to be attacked, when attack meant defeat, he calmly announced his determination to hold his ground as "long as possible."

CONFEDERATE MILITARY MOVEMENTS PRECEDING THE BATTLE.

From their camp at Moody's Spring, where they had arrived Monday night, Generals Price and McCulloch moved forward to the point on Wilson's creek, heretofore described, and went again into camp on the 6th. Scouting parties were at once sent out, especially to discover the Federal position, but with little success, while foraging parties scoured the country in every direction, and were equally inefficient in obtaining information. The combined forces were at once put in position to advance on Springfield, and only awaited the decision of Gen. McCulloch to begin to move. The latter was irresolute and undecided for some days. From the information he possessed as to the strength and character of Lyon's forces and his knowledge of his own, he was fearful of the result of an engagement at that time. He had but little confidence in Price's Missourians, who were somewhat undisciplined and inexperienced, it is true, and at one time he characterized them as "splendid roasting-ear foragers, but poor soldiers."

It is an undoubted fact that at one time Gen. McCulloch had decided *to retreat into Arkansas*. Gen. Price, however, was anxious for an immediate advance and attack. He knew that Lyon's force was inferior even to his own, and that the entire Southern army had but little to risk in offering battle. He knew furthermore, that Lyon ought to be reinforced, and that the chances were that he could and would be, and of course it was desirable that the enemy be attacked before this reinforcement should be effected. The most serious feature considered by McCulloch, that the Missourians were illy disciplined, imperfectly organized, and poorly armed, Price thought would be overcome by their superiority in numbers and *their pluck* in fighting on Missouri soil against a detested enemy—"the Yankee Dutch."

There remains to be shown a good reason why McCulloch did not follow up Lyon and attack him on the 6th; it is true that he gained a victory by waiting, but that victory could have been won four days earlier and made more complete, more decisive and more lasting in its results. And yet McCulloch, on the 8th, seriously meditated a retreat—knowing his enemy's strength as well as his own, and understanding, or supposed to understand, the situation perfectly.

In his report to the Confederate Secretary of War (see Rebellion Records, Series I., Vol. III., p. 745), Gen. McCulloch says: "* * * I asked of the Missourians, owing to their knowledge of the country, some reliable information of the strength and position of the enemy. This they repeatedly promised, but totally failed to furnish, though to urge them to it I then and at subsequent periods *declared I would order the whole army back to Cassville,* rather than bring on an engagement with an unknown enemy. It had no effect, as *we remained four days within ten miles of Springfield, and never learned whether the streets were barricaded, or if any kind of works of defense had been erected by the enemy.*"

Col. Snead says that McCulloch made every effort to discover the condition of Springfield; that he (McCulloch) would frequently sling his rifle over his shoulder, mount his horse and reconnoiter in person; but all to no purpose. Incredible as it may seem, it could not even be ascertained whether or not the Federals had thrown up breastworks, which it might be supposed could be learned from inspection a mile away.

Gen. N. B. Pearce says the first information concerning Gen. Lyon's condition was furnished by two ladies, who, "on a pass to go out of Lyon's lines, came around by Pond Springs, and came to Gen. Price's headquarters and gave the desired information." No corroboration of this story has been obtained, but it is given on the high authority of such a gallant officer and high-minded gentleman as Gen. Pearce, now of Whitesboro, Texas.

At last, Gen. Price lost all patience, and at sunrise on the morning of the 9th, sent Col. Snead over to McCulloch, to say to him that if he did not give orders for an immediate advance he (Price) would resume command of the Missouri State Guard and advance alone, be the consequences what they might. This led to a conference of the general officers at Price's headquarters that afternoon, which conference resulted in orders for an advance on Springfield that very night, the movement to begin at nine o'clock.

GEN. LYON MARCHES OUT TO BATTLE.

Upon the receipt of Gen. Fremont's last message, to the effect that no help would be sent, Gen. Lyon resolved upon attacking his enemy down on Wilson's creek and trusting to the effect of a surprise and a fierce fight. He was led to this course by the fact that he knew his situation would not improve with time, and perhaps by his knowledge of the fact that Price and McCulloch were about to attack *him*.[6] To fight on the defensive about Springfield, with a town full of women and children behind him and an open country adapted to the movements of cavalry, of which he had but a handful, and of which his enemy's force largely consisted, could but result one way—in defeat. The Confederates were expecting to attack, not to be attacked, and if the Federals should fall suddenly upon them it would disconcert them very materially, to say the least. These were the tactics adopted by Gen. Lee when Grant crossed the Rapidan, in the spring of 1864, and by Napoleon, in the first campaign in Italy.

Accordingly, late in the afternoon of the 9th (Friday) word was sent to the subordinate commanders that after nightfall another movement against the Confederates would be made. Between Gens. Lyon and Sweeney, Col. Sigel, and Maj. Sturgis, the plan of attack was agreed upon. That part of the plan which arranged for sending Sigel's brigade around completely to the south and rear of the Confederate position, was, it is said, adopted by Gen. Lyon upon the most urgent suggestions and representations of Col. Sigel himself. The army was to be divided into two columns. The first column, under Lyon, was to consist of three small brigades, the second under Sigel, was to consist of one small brigade composed of two regiments of infantry, two companies of cavalry, and six pieces of artillery.

The first brigade of Lyon's column was composed of three companies of the 1st U. S. regular infantry, as follows: Co. B, Capt. Gilbert; Co. C, Capt.

Plummer; Co. D, Capt. Huston; a company of regular rifle recruits under Lieut. Wood,—the four companies being commanded by Capt. Plummer, of Co. C. Then there were two companies of the 2d Missouri Volunteers, under Maj. P. J. Osterhaus; Capt. Woods' company (mounted) of the 2d Kansas Volunteers; Company B, 1st U. S. regular cavalry, under Lieut. Caulfield, and a light battery of six pieces commanded by Capt. James Totton. The first brigade was commanded by Maj. Sturgis.

The second brigade was commanded by Lieut. Col. Geo. L. Andrews, of the 1st Missouri Volunteers (Blair's regiment), and was composed of the 1st Missouri infantry; Cos. B and E, 2d U. S. regular infantry, under Capt. Fred Steele; one company of regular recruits under Lieut. Lothrop; one company (squad) of mounted recruits under Sergeant Morine, and Lieut. Dubois' light battery of four pieces, one a 12-pounder.

The third brigade was commanded by Gen. Sweeney, and was composed of the 1st Iowa volunteers, under Lieut. Col. Merritt, the colonel, J. F. Bates, being sick in Springfield; the 1st Kansas, under Col. Geo. W. Deitzler; the 2d Kansas, under Col. Mitchell, and about 200 mounted Dade county home guards, under Capt. Clark Wright and Capt. T. A. Switzler.

Gen. Sigel's command consisted of eight companies of the 3d Missouri volunteers (Sigel's regiment), under Lieut. Col. Albert; nine companies of the 5th Missouri, under Col. Salomon; one company, 1st regular cavalry, under Capt. Carr; one company, C, of the 2d U. S. dragoons, under Lieut. Farrand, and six pieces of light artillery manned by details from the infantry recruits under Lieuts. Schaeffer and Schuetzenbach.

THE MARCH BEGUN—ROUTE OF GEN. LYON.

At about 6 P. M. of Friday evening, the 9th, the movement of troops began. Gen. Lyon's column went to the westward, on the Mt. Vernon road, Capt. Gilbert's company of regular infantry having the advance. In a short time it was dark, but the march was continued. Although the march was intended to result in a surprise, and, it was expected, would be conducted silently, yet there was a great deal of noise made. The Iowa and Kansas volunteers were disposed to exercise their vocal organs, and camp songs of all sorts were sung *con spirito*, along the march. The 1st Iowa had a favorite song, the burden of which ran:—

> So let the wide world wag as it will,
>
> We'll be gay and happy still.
>
> Gay and happy, gay and happy,
>
> We'll be gay and happy still.

The strains of this song were wafted out over the prairie, loud enough, it would have seemed, to have been heard by McCulloch's pickets, if any were out. The Kansas men sang the "Happy Land of Canaan," and raised the neighborhood with their vocal efforts. Toward midnight, however, the line became more quiet, by Gen. Lyon's orders. The latter had remarked during the march that the Iowa troops had too much levity in their composition to do good fighting, but added that he would give them an opportunity to show what they were made of. It so turned out that the general was mistaken in his estimate of the fighting qualities of the Hawkeyes.

Lyon marched west from Springfield on the Mt. Vernon road, about five miles, or a little east of where the town of Brooklyn now stands, when he turned south, and made his way over neighboring roads and across prairies as best he could nearly six miles, when he reached a point within striking distance of Price's Missourians. The center of the camp of the Southerners was about six miles west, and about seven miles south of the public square of Springfield. Gen. Lyon had for guides Pleasant Hart, Parker Cox, and other men. Nearly twenty men have come forward to claim this distinction.

It was 1 o'clock in the morning when the advance discovered the camp-fires of the Missourians. The command was then halted, and the ground reconnoitered as well as possible until the dawn of day, when it again moved forward and formed a battle line, moving a little southeast so as to strike the extreme northern point of the enemy's camp.

COL. SIGEL'S ADVANCE.

Sigel left "Camp Fremont," on the south side of Springfield, at about 6:30, p. m., taking at first the "wire" road, or road to Cassville and Fayetteville, along which the telegraph wire ran. About four miles southwest of town, the command left the main Cassville road, which led directly through McCulloch's camp, and bore south, and then along a road parallel with the Cassville road, and in the same general direction, until below the Christian county line. Col. Sigel had for guides, C. B. Owen, John Steele, Andrew Adams, Sam (or Jo.) Carthal and L. A. D. Crenshaw. Sigel's column marched perhaps twelve or thirteen miles, passing clear around the extreme southeastern camp of the enemy, and arriving at daylight within a mile of the main camp. Taking forward the two cavalry companies of Carr and Farrand, Col. Sigel contrived to cut off about forty men of McCulloch's troops, who had gone out early to forage, and were engaged in digging potatoes, picking roasting ears, gathering tomatoes and procuring other supplies for their individual commissary departments. These captures were made in such a manner that no news of the Federal advance from this quarter was brought into the Confederate camp. Moving cautiously up, Sigel planted four pieces of his artillery on a little hill, in plain view of the Confederate tents, which spread out to his front and right. The two regiments of infantry advanced so as to command the Fayetteville road at the point where it crosses Wilson's creek, while the two companies of cavalry guarded the flanks. In this position the command rested, awaiting the sound of Lyon's gun as a signal to open the ball. The prisoners were left in charge of Capt. Flagg, with his company (K) of the 5th Missouri.

In conformity to the plan agreed upon between the Federal commanders, Sigel disposed his troops so as to command the Fayetteville road, and prevent the Confederates from retreating by that thoroughfare. It is claimed by officers of both armies that, had an avenue of retreat been left open, it is highly probable that the result of the day's battle would have been different.

Lyon had left behind him the Greene and Christian County Home Guards to take care of Springfield, directing the officers in command to watch the Fayetteville road below where Sigel turned off, and send word to him across the country, should the Confederates be found approaching from that quarter. This is a circumstance corroborative of the theory that Lyon knew that the Confederates meditated a night attack on him (as they did) or believed that such was a fact. Everything in Springfield had been gotten ready for a retreat. Wagons were loaded, and the funds of the bank were secured for transfer, and were being guarded by the Home Guards. The citizens were in quite a state, to be sure.

LYON OPENS THE BATTLE.

In describing the battle of Wilson's Creek in this history, which, it is believed, contains the only fully elaborate and accurate account ever published, of that memorable contest, it is proper to do so in detail. The statements herein made have been derived from the official reports of commanders, and from the fairest accounts of actual participants. Care has been taken to discard all reports which are highly colored, sensational, not corroborated by undisputed facts, and savoring of the improbable. Both Federal and Confederate accounts of this character have been rejected. The Federal accounts believed to be the most reliable are those furnished by Maj. (General) Sturgis, Lieut. Col. Merritt of the 1st Iowa; Lieut. Col. Blair and Maj. Cloud, of the 2d Kansas; Maj. J. M. Schofield, then of the 1st Missouri; Capt. Totten and Lieut. Dubois of the artillery, and Capt. Steele of the regulars; Capt. Wright of the Home Guards, all of Lyon's column; and Gen. Sigel, Dr. S. H. Melcher, the guides, and Capt. Carr, of Sigel's column. The Confederate or Southern accounts relied upon, are the official reports of Gens. Price, McCulloch, Pearce, Clark, Rains, McBride and Parsons; reports of Col. John T. Hughes, of Slack's division, and Col. John R. Graves, of Rains' division; letters from Col. Thos. L. Snead, Asst. Adj. Gen. of Gen. Price, and Lieut. W. P. Barlow, of Guibor's battery; reports of and letters from Col. T. J. Churchill, 1st Arkansas Mounted Riflemen; Col. James McIntosh, and Lieut. Col. B. T. Embry, 2d Arkansas Mounted Riflemen; Lieut. Col. D. McRae, of McRae's battalion, Arkansas Volunteers; Col. Lewis Hebert, Lieut. Col. S. M. Hyams and Maj. W. F. Tunnard, 3d Louisiana Volunteers; Col. E. Greer, South Kansas-Texas Regiment Cavalry; Capt. J. G. Reid, of Reid's Arkansas Battery; Col. John R. Gratiot, 3d Arkansas; Col. J. D. Walker, 4th Arkansas; Col. Tom P. Dockery, 5th Arkansas Infantry; Col. De Rosey Carroll, 1st Arkansas Cavalry, and other commissioned officers, and many private soldiers and a few citizens.

Maj. Sturgis, who assumed command of Lyon's column after the battle, states that at daylight, Lyon's battle line was formed, the infantry in front, closely followed by Totten's battery, which was supported by a reserve. In this order the line advanced but a few hundred yards, when the first outpost of Price's men was encountered. Firing was commenced instantly, and the outpost hurriedly retreated. This was the advance of Rains' division. The Federal line then halted, and Capt. Plummer's battalion of regulars, with the Dade County Home Guards on his left, was sent to the east across Wilson's creek, and ordered to move toward the front, keeping pace with the advance on the Federal left. The main line then swept forward, and after crossing a considerable ravine and ascending a high ridge, a full view of a line of Rains' skirmishers was had. Maj. Osterhaus' two companies of the 2d Missouri, and two companies of the 1st Missouri, under Capts. Yates and John S. Cavender, were deployed to the left, all as skirmishers. Firing between the two skirmish lines now became very severe, and Totten's battery, then in position, opened with shell, and the boom of the cannon and the crashing of the bombs added to the excitement.[7]

The 1st Missouri, Col. Andrews, and the 1st Kansas, Col. Dietzler, were now hastily moved to the front, supported by Totten's battery; the 2d Kansas, Col. Mitchell, Steele's battalion, and Dubois' battery, were held in reserve. The 1st Missouri took its position in front, upon the crest of a small elevated plateau. The 1st Kansas went to the left of the 1st Missouri, while Totten's battery was placed opposite the interval between the two regiments. Osterhaus' two companies occupied the extreme right, with their right resting on a ravine, which turned abruptly to the right and rear. Dubois' battery, supported by Steele's battalion, was placed seventy-five yards to the left and rear of Totten's guns, so as to bear upon a well-served Confederate battery (believed to have been Capt. Woodruff's "Pulaski Artillery," of Arkansas), which had come into position to the left and front on the opposite side of Wilson's Creek, and was sweeping with canister the entire plateau upon which the Federals were posted.

The Missourians now rallied in considerable force under cover at the foot of the slope and along it in front and opposite the Federal right, toward the crest of the main ridge running parallel to the creek. During this time Plummer's battalion had advanced along the ridge about 500 yards to the left of the main Federal position, and had reached the terminus of this ridge, when he found his further progress arrested by a force of infantry (a portion of McCulloch's division), which was occupying a cornfield (Mr. Ray's) in the valley. At this moment the "bang" of a cannon was heard more than a mile to the south, at about the point where Sigel was supposed to be. This fire was apparently answered from the opposite side of the valley, at a still greater distance, the line of fire of the two batteries being apparently east and west, and nearly perpendicular to Totten's and Dubois' batteries. After about ten or twelve shots this firing ceased, and nothing more was heard of Sigel until about 8:30, when a brisk cannonading was heard for a few minutes, about a mile to the right of that heard before, and still further to the rear.

Early in the engagement the 1st Iowa had been brought up from the reserve to the front, and immediately became hotly engaged, doing good fighting and winning the praise of Gen. Lyon, who thought at one time that men who sang rollicking songs would not fight well.

The entire Federal line was now successfully advanced with much energy, and apparently with every prospect of success. The firing, which had been spirited for half an hour, now increased to a continuous roar, heard miles away—in Springfield, plainly. Capt. Totten's battery came into action by section and by piece, as the nature of the ground would admit, it being wooded, with much black-jack undergrowth, and played vigorously upon the Confederate lines with considerable effect.

More desperate fighting was not done during the civil war. The men of the West were fighting. For fully half an hour the armies fought over the hill before described—"Bloody Hill," it was afterward called. The 1st Kansas gave way and went to the rear, but the 1st Iowa promptly took its place, and the fighting went on. Back and forth over the ground they went. Now the Union troops fell back a few yards, then advanced again and drove the secession troops a short distance, then the latter advanced, and so it was for half an hour. At last the Federals were left in possession of the ground for a short time, the Confederates falling back and reforming.

Meantime Plummer's battalion on the Federal left had encountered McIntosh's regiment of Arkansas riflemen, and Hebert's 3d Louisiana regiment, in Ray's cornfield and been driven back with considerable loss. The Arkansas and Louisiana regiments both belonged to McCulloch's army. They would have annihilated Plummer almost, but just as they were preparing to do so Dubois' battery opened with shells, filling the cornfield full of them, and making it untenable for any troops, and the two regiments retreated in some disorder. Steele's battalion was supporting Dubois' battery on this occasion. Plummer was severely wounded.

Just now there was a momentary cessation of firing, the advantage being with the Federals, and it became apparent that some of the Southerners desired to retreat, but they soon learned that they were practically surrounded, for there was no road to the east or the west, and the only outlet from their position, the Fayetteville road, was held by Sigel. The only way therefore to get out was to fight out. Quite a number of the Missourians were in confusion. Their horses were frightened and became uncontrollable, and the men galloped about aimlessly, and wildly. Some of them got away from the battle field and rode away to Cassville panic-stricken and reporting that Gen. Price's army had been "all cut to pieces" by an overwhelming force of Federals! The greater portion of Lyon's line was quiet for a time, and some thought the victory had been won.

Along the right of the Federal line, however, the 1st Missouri was hotly engaged with McBride's division of Missourians and was about to be overcome. Lyon hurried the 2d Kansas to its relief and saved it. During the temporary lull in the firing the Federal line was reformed under the direction of Lyon himself. Steele's battalion, which had been supporting Dubois' guns, was brought forward to the support of Totten's, and preparations were made to withstand another attack, which, as could be ascertained by the shouts of the enemy's officers, plainly audible, was being organized.

Scarcely had Lyon disposed his men to receive the attack when his enemy again appeared with a very large force along his entire front and moving toward his flanks as well. At once the firing again began and for a time was inconceivably fierce along the entire line. The Confederates were in three lines *in some places* the front line lying down, the second kneeling, the third line standing, and all the lines and every man loading and firing as rapidly as possible. Every available Federal battalion was now brought into action, and the battle raged with great fury for an hour, the scales seeming all the time nearly equally balanced, sometimes the Federal troops and then the Confederates gaining ground and then losing it, while all of the time some of the best blood in the land was being spilled as recklessly as if it were ditch-water.

How they did fight, these men of both armies!—fought until their gun-barrels became so hot they could scarcely hold them—fought when their leaders fell and without commands—fought when the blood and brains of their comrades were spattered into their faces—fought, many of them, until they died. By and by, as the Confederate fire never slackened, but was constantly increased by the arrival of reinforcements, and as some of the Federals reported that their cartridges had given out, detachments of the latter began to give way, and Gen. Sweeney and Gen. Lyon were engaged from time to time in bringing them back into the fight.

DEATH OF GEN. LYON.

Early in this engagement, while Gen. Lyon was walking and leading his horse along the line on the left of Totten's battery, his horse, the iron gray, was killed and he was wounded in two places, in the head and in the leg. Captain Herron, of the 1st Iowa,[8] states that he saw the horse fall, and that the animal sank down as if vitally struck, neither plunging nor rearing. Lyon then walked on, waving his sword and hallooing. He was limping for he had been wounded in the leg. He carried his hat, a drab felt, in his hand and looked white and dazed. Suddenly blood appeared on the side of his head and began to run down his cheek. He stood a moment and then walked slowly to the rear. Capt. Herron states that he was within twenty feet of Lyon when this happened, near enough to observe that he was wearing his old uniform, that of captain in the regular army.

When he reached a position a little in the rear Lyon sat down and an officer bound a handkerchief about his wounded head. He remarked despondingly to Maj. Schofield, of Blair's regiment, one of his staff: "It is as I expected; I am afraid the day is lost." The Major replied: "O, no, General; let us try once more." Major Sturgis then dismounted one of his own orderlies and offered the horse to Lyon, who at first declined the animal, saying: "I do not need a horse." He then stood up and ordered Sturgis to rally a portion of the 1st Iowa which had broken. Sturgis, in executing this order, went to some distance from his general. The 1st Iowa was being ordered forward by a staff officer, when some of the men called out, "We have no leader," "Give us a leader, then," etc. Lyon immediately asked to be helped on the orderly's horse. As he straightened himself in the saddle the blood was dripping off his heel from his wounded leg. Gen. Sweeney rode up and Lyon spoke quickly to him, "Sweeney, lead those troops forward (indicating the 1st Iowa) and we will make one more charge."

Then, swinging his hat, Lyon called out to the 2d Kansas regiment, "Come on, my brave boys, (or 'my bully boys,' as some say), I will lead you; *forward*!" He had gone but a few yards when he was shot through the body. One of his orderlies, a private named Ed. Lehman, of Co. B, 1st U. S. cavalry, caught him in his arms and lowered him to the ground. With the breath still feeling at his lips, and his great heart throbbing and striking his own death-knell, the dying chieftain gasped, "Lehman, I'm going," and so passed away his spirit through the battle-clouds to the realms where is everlasting peace. The place where Lyon fell was afterward called "Bloody Point." A heap of stones marks the spot to this day. Lyon's body was borne to the rear by Lieut. Schreyer, of Capt. Tholen's company of the 2d Kansas, assisted by Lehman and another soldier.

STILL THE BATTLE GOES ON.

In the meantime the disordered Federal line was rallied and reformed. The 1st Iowa took its place in the front, and Major Sturgis says, "fought like old veterans." The Kansans and the Missourians were also doing well, and the Confederates were driven back, only to come again. The situation of the Federals was now desperate. The commander, Gen. Lyon, was killed; Gen. Sweeney was wounded, Col. Deitzler, of the 1st Kansas, lay with two bullets in his body; Col. Mitchell, of the 2d Kansas, by the same fire that killed Lyon, was severely wounded (it was thought at first mortally) and as he was borne from the field called to an officer of Maj. Sturgis' staff, "For God's sake support my regiment;" Col. Andrews, of the 1st Missouri, and Col. Merritt, of the 1st Iowa, were wounded; and thus it was that all of the regimental commanders of Lyon's column were wounded. Still the battle went on.

THE LAST GRAND CHARGE OF PRICE'S MEN.

The great questions in the minds of Sturgis and Sweeney and the other Federal officers, who had been informed of the plan of attack agreed upon were, "Where is Sigel? Why doesn't he co-operate?" Although it seemed as if there must be a retreat should the Southerners make another vigorous charge, yet if Sigel should come up with his near 1,000 men, and make an attack on Price's right flank and rear, then the Federals could go forward with strong hopes of success. If Sigel had been whipped, however, there was nothing left but to retreat.

Maj. Schofield, Lyon's chief of staff, rode to Sturgis and informed him that Lyon was killed and Sigel could not be heard from, and moreover, that the ammunition was about exhausted, some of the troops being entirely out. Sturgis thereupon assumed command—although only a major at the time. He at once summoned the principal officers left and consulted with them. All agreed that unless Sigel made his appearance very soon there was nothing left but to retreat, if indeed retreat were possible.

The consultation was brought to a close by the advance of a heavy column of infantry from towards the hill where Sigel's battery had been heard at the beginning of the struggle. These troops carried flags which, drooping about the staffs, much resembled the stars and stripes, and Sturgis and Schofield say the troops had the appearance of Sigel's. A staff officer in front of where the consultation was going on rode back and called out delightedly *"Yonder comes Sigel! Yonder comes Sigel!"* and the officers departed, each to his command to arrange for the expected change in the programme.

On came the moving mass in Sturgis' front, the soldiers cool and steady as grenadiers. Down the hill across the hollow in front they swept and took position along the foot of the ridge on which the Federals were posted. And now, "they are rebels!" was heard from the more advanced of the Kansans and Iowans. Suddenly a battery (Guibor's) which had followed the line and had reached the hill in front of "Bloody Hill," wheeled about, unlimbered and the command *"Fire!"* rang out and the guns belched forth shrapnel and canister before the trail pieces had hardly touched the ground. The infantry at the foot of the hill, now began firing and slowly ascending the hill, and at once commenced the fiercest and most bloody struggle of all that bloody day.

Lieut. Dubois' battery, on the Federal left supported by Osterhaus' two companies and the rallied fragments of the Missouri 1st, opened on the new battery (Guibor's) and soon checked it. Totten's battery, still in the Federal center, supported by the Iowans and regulars, seemed to be the main point of the Confederate attack.

The Missourians frequently came up within twenty feet of the muzzles of Totten's guns and received their charges of canister full in their faces, and the two clouds of battle smoke mingled until they seemed as one.

For the first time during the day the Federal line never wavered and the Confederate line never flinched. At one time Capt. Steele's battalion, which was some yards in front, together with the left flanks, was in danger of being overwhelmed and captured, the contending lines standing so close that the muzzles of their guns almost touched. Capt. Granger, of Sturgis' staff, ran to the rear and brought up the supports of Dubois' battery, consisting of Osterhaus' battalion, detachment, of the 1st Missouri, 1st Kansas, and two companies of the 1st Iowa, in quick time, and took position on the left flank, and poured in a heavy volley upon the Confederates, which was so murderous and destructive that that portion of the line gave way. Capts. Patrick E. Burke and Madison Miller, and Adjutant Hiscock, of the 1st Missouri, were especially mentioned for gallantry in this assault.

The entire Confederate line now fell back a short distance and began again forming. Sturgis took advantage of this lull in the storm to make good his retreat. Perceiving that Totten's battery and Steele's battalion were entirely safe, for the present, and directing Capt. Totten to replace his disabled horses as soon as possible, Sturgis sent Dubois' battery to the rear with its supports to take up a position on the hill in the rear and cover the retreat. The 2d Kansas, on the extreme right, having been nearly out of ammunition for some time, was ordered to withdraw, which it did bringing off its wounded. This, however, left the Federal right flank exposed, and the Missourians at that point, to the number of 100 or more, advanced at once; they were driven back, however, by Steele's battalion of regulars and joined the main force reforming in the rear.

RETREAT.

Maj. Sturgis gave the order to retreat as soon as his enemy had fallen back and enabled him to do so. Totten's battery, as soon as his disabled horses could be replaced, retired with the main body of the infantry, while Capt. Steele met the feeble demonstrations of a few plucky Missouri skirmishers who had not fallen back with the main line and were picking away at the Federal right flank. The whole Federal column now moved unmolested and in tolerable order to the high open prairie east of Ross' spring and about two miles from the battle ground. The artillery and the ambulances, were brought off in safety. After making a short halt on the prairie the retreat was continued to Springfield over substantially the same route taken to the field.

Just after the order to retire had been given, and while Sturgis was undecided whether to retreat from the field entirely or take up another position, one of Sigel's non-commissioned officers (Sergt. Frœlich) arrived on a foam-covered horse and reported that Col. Sigel's brigade had been totally routed, his artillery captured, and the colonel himself either killed or taken prisoner.

On reaching the Little York road Sturgis encountered Lieut. Farrand, with his company of dragoons, one piece of artillery and a considerable portion of the 3d and 5th Missouri, all of Sigel's command, which had made their way across the country in order to unite with the main command and be saved from entire destruction. The march was resumed, but the command did not succeed in reaching Springfield until five o'clock in the evening.

Lyon's column began the attack at about 5 in the morning and it was half-past 11 when the battle ended; the main body of the troops were engaged about six hours.

SIGEL'S PART IN THE FIGHT.

It is proper now to consider the part taken by Col. Sigel and his brigade in the battle of Wilson's Creek. It has been stated that he had moved entirely around the southern end of the Confederate line of camp, and on a previous page we left him with his guns "in battery," and his infantry and cavalry in line commanding the Fayetteville road, and ready to open fire as soon an the sound of Lyon's guns could be heard up the valley, nearly two miles.

At 5:30, early in the morning, the rattle of musketry was heard, apparently nearly two miles away, to the northwest. *"Bang! Bang! Bang! Bang!"* in rapid succession, went the four guns of Lieuts. Schaeffer and Schuetzenbach, as they discharged their contents into and among the tents of McCulloch's camp. A few more rounds and the Confederates abandoned their tents and retired in haste toward the northeast and northwest. This fighting was done just across the line, in Christian county, on Sharp's farm, which runs up to the county line, on which stands Mr. Sharp's house.

McCulloch's troops, infantry and cavalry, soon began to form, and Sigel brought forward his entire line into and across the valley, the two companies of cavalry to the right, the artillery in the center and the infantry on the left. After a period of irregular firing for about half an hour, the Confederates retired into the woods and up the adjoining hills. The firing toward the northwest was now more distinct, and it was evident that Gen. Lyon had engaged the enemy along the whole line. To give assistance to him—to be able to co-operate with him if necessary, and to drive the enemy in his own front, Sigel again advanced, this time toward the northwest, intending to attack the Confederates in the rear.

Marching forward, Sigel struck the Cassville road, making his way through a number of cattle and horses, and arriving at an eminence, which had been used as a slaughter-yard by McCulloch's men. This was on Sharp's farm and near the house. At and near Sharp's house, on the road, some of McCulloch's men, who were straggling back from the fight in front, came unawares on Sigel's men and were taken in. Sigel, after a brief conference with some of his officers, at once concluded that Lyon had been successful and was driving the Confederates before him. Knowing that this was the only avenue of retreat left open, and imagining that here was a grand opportunity for stopping it up and bagging several thousand "rebels," the colonel hurriedly formed his troops across the road, planting the artillery in the center on the plateau, and a regiment of infantry, and a company of cavalry on either flank, and awaited the coming of what seemed to him to be the *vanquished* Confederates, large numbers of whom could be seen moving toward the south along the ridge of a hill about 700 yards opposite the right of the Federal right.

It was now about half-past eight o'clock, and the firing in the northwest, where Lyon was supposed to be, and where he really was fighting, had almost entirely ceased. At this instant, Dr. S. H. Melcher, the assistant surgeon of Salomon's regiment, and some of the skirmishers came back from the front, where desultory firing had been going on, and reported that Lyon's men were coming up the road, for they could be seen plainly, and the gray-coated Iowa regiment plainly distinguished. At once, Lieut. Col. Albert, of the 3d Missouri, and Col. Salomon, of the 5th, notified their regiments *not to fire* on the troops coming in this direction, for they were friends, and Sigel himself gave the same caution to the artillery.

Everybody was surprised at this unexpected turn of affairs, and the Germans of Sigel's and Salmon's regiments began jabbering away delightedly, and the color-bearers were beckoning with their flags to the advancing hosts to "come on"—when, all at once, two batteries of artillery, one on the Fayetteville road and one on the hill where it was supposed Lyon's men were in pursuit of the flying Confederates, opened with cannister, shell and shrapnel, while the gray-coated troops, supposed to be the Iowans, advanced from the Fayetteville road and attacked the Federal right, and a battalion of cavalry made its appearance, apparently ready and waiting to charge!

The jabbering of the German soldiers was now something wonderful, but it had a different tone from that of a few minutes previously! It is impossible to describe the consternation and frightful confusion that resulted. So surprised and frightened were the soldiers that they could not understand these were *Confederates* who were firing upon them and coming rapidly forward to sweep them from the face of the earth. They hurried and skurried about crying, some in English: "It is Totten's battery!" others in German: *"Sie haben gegen uns geschossen! Sie irrten sich!"* (They are firing against us! They make a mistake!) And then making no effort to fight worthy of the name, they began to retreat.

The artillerymen, all of whom were recruits from the infantry, who had seen but little service of any kind, could hardly be brought forward to serve their pieces, although directed by Sigel himself; the infantry would not level their guns until it was too late; indeed, they could not be made to stop running, let alone to turn and fight. Salomon cursed in German, in English, in French. Sigel threatened and bullied and coaxed. No use. As well try to stop a herd of stampeded buffaloes. Lieut. Farrand, with his company of cavalry brought off one piece of artillery which had not been unlimbered and put in position, and away it went the wheels bouncing two feet from the ground and the postilions lashing their horses like race riders.

On came McCulloch's and Price's men, the Louisiana regiment of Col. Hebert (pronounced Hebare) which had been mistaken for the 1st Iowa because of its pretty steel gray uniform, was in front, and following them were the Arkansas regiments of Dockery and Gratiot, the 5th and 3d, Greer's regiment of Texas cavalry, Lieut. Col. Major's Howard and Chariton county battalion, Johnson's battalion mounted Missourians, and some other detachments. Up to the very muzzles of the cannons they came, killing the artillery horses and what artillerymen were reckless enough to remain, firing fairly into the faces of the panicky Teutons and forcing them to throw themselves into the bushes, into by-roads, anywhere to escape and scamper away as fast as their legs could carry them. The color-bearer of Sigel's own regiment was badly wounded; his substitute was killed, and the flag itself was captured by Capt. Tom Staples, a Missourian, of Arrow Rock, Saline county.

When the plateau was reached, the cannon captured and the field gained, the infantry stopped and cheered, Reid's and Bledsoe's batteries fired parting salutes into the flying blue-coats, and then, leaving the cavalry to pursue, both infantry and artillery turned about and went up to the other end of the valley to assist their brethren in that quarter, and to participate in the final triumph of the day.

Away went the Germans, down to the south into Christian county, throwing away guns, cartridge boxes, even canteens,—everything that hindered rapid flight,—wandering about and hiding when they could with the Texas, Arkansas, and Missouri cavalry leaping upon them incessantly and slaying them wherever they made the least show of resistance. At Nowlan's mill, on the James, three miles from the battle-ground, it was told that four fugitives skulked under the mill-dam and, refusing to come out, were riddled with buckshot.

The next day men lay scattered all over the country, wounded or dead; and yet Sigel lost but comparatively few killed. Prisoners were taken in great numbers—run down by the Texas rangers and driven in like flocks of sheep, as timid now and as harmless. Sigel himself got panicky after awhile and fled for Springfield, across the country, accompanied by only two guards, giving rise to the wicked stanza of the song sung in the Confederate camps after the battle, concerning the battle of Wilson's Creek,—how,

> Old Sigel fought some on that day,
>
> But lost his army in the fray;
>
> Then off to Springfield he did run,
>
> With two Dutch guards, and nary gun.

At Mrs. Chambers' house, four miles south of Springfield, Col. Sigel and his two guards halted and procured a drink of water, and then rode away to Springfield, as rapidly as their jaded horses could carry them. Sigel himself arrived at Springfield with but one orderly.

Only the cavalry under Carr and Farrand, the one piece of artillery, two caissons and about 150 infantry came off in anything like order, and these followed down the wire road some miles to the west and then turned off due north and united with Sturgis' column, near the Little York road. Only four pieces of artillery were captured at the time of the charge on the hill, for those were all that were in position. The two others were in the rear. In attempting to get one of them away a wheel horse was killed, and the drivers abandoned the gun, after first spiking it as best they could. The gun that was saved was first abandoned out on the Fayetteville road, and hauled off at first by hand a short distance, Capt. Flagg employing the prisoners and soldiers as artillery horses.

Concerning the retreat of that portion of Sigel's force which went to the westward, Lieut. Chas. E. Farrand (then of the Second Regular Infantry) commanding the company of cavalry before mentioned, writes:—

> Upon finding myself with my company alone, I retired in a southerly direction, and accidentally meeting one of the guides (Mr. Crenshaw), who had been employed in taking us to the enemy's camp, I forcibly detained him until I could collect some of the troops, whom I found scattered and apparently lost. I halted my company and got quite a number together, and directed the guide to proceed to Springfield, via Little York. After proceeding a short distance, we came upon one of the pieces which had been taken from Col. Sigel. Although the tongue of the limber was broken, one horse gone, and one of the remaining three badly wounded, we succeeded in moving it on. Some distance in advance of this we found a caisson, also belonging to Col. Sigel's battery. I then had with me Sergt. Bradburn, of company D, 1st cavalry, and Corporal Lewis and Private Smith, of my

own company (C, 2d dragoons). My company being some distance in advance, I caused the caisson to be opened, and on discovering that it was full of ammunition, I determined to take it on. I and the three men with me tried to prevail upon some of the Germans to assist us in clearing some of the wounded horses from the harness, but they would not stop. After considerable trouble, my small party succeeded in clearing the wounded horses from the harness, hitching in two more and a pair of small mules I obtained, and moved on, Corporal Lewis and Private Smith driving, while Sergt. Bradburn and I led the horses. After reaching the retreating troops again I put two other men on the animals, and joined my company with my three men. Before reaching Springfield it became necessary to abandon the caisson,[9] in order to hitch the animals to the piece. This was done after destroying the ammunition it contained. Lieut. Morris, adjutant to Col. Sigel's command, assisted me in procuring wagons, which we sent back on the road after the wounded.

The route of retreat taken by Lieut. Farrand and Capt. Flagg, and the fragments of Sigel's command, 400 in all, was down the wire road a short distance, and then north to the Mt. Vernon road. While marching northward this body of disordered men was only within two or three miles of the entire Southern army for three or four hours. Why Generals Price and McCulloch did not send out a small force of mounted men and take prisoner every man, which could very easily have been done, is inexcusable, certainly.

DR. S. H. MELCHER'S ACCOUNT.

Mention has been made of Dr. Samuel H. Melcher, who as assistant surgeon of Col. Salomon's 5th Missouri (Dr. E. C. Franklin, being surgeon), was present at the battle of Wilson's Creek with Sigel's command. To the writer hereof Dr. Melcher, now of Chicago, sends his recollections of the events of the memorable contest. After narrating the preliminary movements of Sigel, substantially as heretofore given, Dr. Melcher says:—

> * * * Gen. Sigel soon gave the order to fire, which was responded to with rapidity, but our guns being on an elevation, and the Confederates being in a field which sloped toward the creek, the shots passed over their heads, creating a stampede but doing little, if any, damage to life or limb. In vain I and others urged the artillerymen to depress the guns. Either from inability to understand English, or, in the excitement, thinking it was only necessary to load

and fire, they kept banging away till the whole camp was deserted. * * * The command then moved on till it reached the Fayetteville road and Sharp's house. While the command was taking position, I with my orderly, Frank Ackoff, 5th Missouri, went into the abandoned Arkansas camp where I found a good breakfast of coffee, biscuit and fried green corn. * * * Most of the tents were open—a musket with fixed bayonet being forced into the ground, but up, and the flap of the tent held open by being caught in the flint lock. At that time, besides a few Confederate sick, there were in the camp Lieut. Chas. E. Farrand, in command of the dragoons, and his orderly. Half an hour later, some straggling parties from the 3d and 5th Missouri, set fire to some wagons and camp equipage.

* * * The four guns were in the front, supported by the 3d Missouri, with the cavalry and dragoons on the left in the timber. The 5th Missouri was in reserve, except Co. K, Capt. Sam'l A. Flagg, which was further in the rear, guarding some thirty or forty prisoners. At this time, scattering shots were heard at some distance in our front, but no heavy firing. Armed men, mostly mounted, were seen moving on our right in the edge of the timber. * * *

It was smoky, and objects at a distance could not be seen very distinctly. Being at some distance in front of the command, I saw a body of men moving down the valley toward us, from the direction we last heard Gen. Lyon's guns. I rode back, and reported to Gen. Sigel that troops were coming, saying to him, "They look like the 1st Missouri." [Iowa?] They seemed moving in a column. * * * By this time, Sigel could see them. Not seeing their colors, I suggested to Sigel that he had better show his, so that if it was our men they might not mistake us—Sigel's brigade not being in regulation uniform. Gen. Sigel turned and said: "Color-bearer, advance with your colors, and wave them—wave them three times." As this order was being obeyed, Lieut. Farrand, with his orderly, arrived from the Arkansas camp, each bearing a rebel guidon, which they had found, and with which they rode from the right of the line, near Sharp's house, directly in front of the color-bearer of Sigel's regiment. Then there was music in the air. A battery we could not see opened with grape, making a great deal of noise as the shot struck the fence and trees, but not doing much damage, as far as observed, except to scare the men, who hunted for cover like a flock of young partridges, suddenly disturbed. The confusion was very great, many of the men saying, "It is Totten's battery! It is Totten's battery!" The impression seemed to be general that Totten was firing into us, after seeing the rebel guidons of Farrand, as it was the common understanding that the Confederates had no grape, and these were grape shot, certainly.[10]

Gen. Sigel now evidently thought of retreat, as the only words I heard from him were, "Where's my guides?" [Instances of individual cowardice among Sigel's officers are here given.] I assisted Lieut. Emile Thomas (now of St. Louis), the only officer of his company that had the grit to stay, to reform the men. I do not know if we could have succeeded, had not a Confederate cavalry battalion suddenly appeared in our front, on the line of retreat. For a moment the two commands gazed upon each other, and then came a terrible rattle of musketry, and a great hubbub and confusion in the direction of Sigel's command, which was just around a bend in the road to our rear.

In a twinkling, men, horses, wagons, guns, all enveloped in a cloud of dust, rushed toward us, and in spite of Lieut. Thomas's utmost efforts, Company F started with all speed down the Fayetteville road toward the Confederate cavalry. The latter, seeming to think that they were being charged upon, wheeled and got out of the way very quickly! The bulk of Gen. Sigel's command turned to the east and were followed by a Confederate command, that captured one gun at the creek, many prisoners, and left a considerable number of killed and wounded along the road.

Perhaps one-third of the command went southwest, and halted at the next house beyond Sharps' on the Fayetteville road, and here Dr. Smith, who was Gen. Rains' division surgeon, came up, with a long train of wagons and coaches, and was captured, but at once released on my intervention. [After this, Dr. Melcher accompanied Dr. Smith to the battle-field.] * * * The one gun that was abandoned on the Fayetteville road was really saved by Capt. Flagg, whose men drew the gun by hand till they found some horses, and the Confederate prisoners carried the ammunition in their arms. * * * They came into Springfield the same evening by way of Little York.

Sigel's reasons for his defeat must here be given. He states that he tried to obey his orders to attack the enemy in the rear and to cut off his retreat. This he did, but he also cut off his own retreat very nearly, a circumstance he had not counted upon.

The time of service of one of his two regiments of infantry, the 5th Missouri, Salomon's, had expired some days before the battle and they had clamored to go home. On the first of August he had induced them to remain with the army eight days more. This latter term had expired the day before the battle. The men therefore were under no obligations to fight, except that they had marched out to do so, and when the time came, suddenly remembered that "they did not have to fight." The 3d regiment, Sigel's own, was not the old 3d, that fought at Carthage; that regiment, its time having expired, had been mustered out, and the new regiment was composed of 400 new recruits and of but a few other men who had seen service. The men serving the artillery were new recruits who knew next to nothing of gunnery, and were commanded by two lieutenants whose only experience as artillerists had been in the Prussian army in a time of peace. Again it is stated that only about half of the companies were officered by men with commissions, which Sigel says, was the fault of the three months' service.

But over all it is claimed that Sigel's complete defeat was the result of an attack of vastly superior forces, the flower of McCulloch's army, that was permitted to approach fatally near under the mistake that they were friends instead of enemies.

As explaining and detailing something of the retreat of that wing of Sigel's command which turned to the east, the following statement of Captain (now General) E. A. Carr, who, as previously stated, commanded the advance guard of Sigel's brigade, may be found of interest:—

"At about 9 o'clock Capt. Carr received word that Sigel's infantry were in full flight and that he was to retreat with all haste. After galloping away as best he could for about a mile and a half to the rear, Carr came upon Sigel at the spring where the army had halted the first night when returning from Dug Spring some days before. After a brief consultation it was decided to move south on the Fayetteville road until there was a chance to go out and circle around the pursuing enemy and then strike for Springfield. There were then present at the spring Sigel, Carr, Lieut.-Col. Albert, Carr's 56 cavalry, 200 of Sigel's badly demoralized infantry, one piece of artillery, and two caissons. After "retiring" rather hastily for a mile or so a body of cavalry was observed in front, and Sigel sent Carr up to see the condition of affairs and report at once. Arriving at the front Carr discovered that the Confederate cavalry were coming in from the right and forming across the road, to stop the retreating Federals and send them back to the care of McCulloch's division again. Reporting at once to Sigel, that officer directed Carr to turn off at the first right-hand road, which happened to be near the point where he (Carr) then stood. Retreating along this road in a brisk walk Sigel asked Carr to march slowly so that the footmen could keep up. Carr replied that unless they hurried forward they would be cut off at the crossing of Wilson's Creek, and that the infantry ought to march as fast under the circumstances as a horse could walk. Sigel then said, "Go on, and we will keep up." On arriving at the creek, however, and looking back, Carr saw that the infantry had not kept up, but that a large body of Texas and Arkansas cavalry was moving down and would form an unpleasant junction with him in a few seconds. "To use a Westernism," says Gen. Carr, "there was no time for fooling then, and as I had waited long enough on the slow-motioned infantry to water my horses, and they were not yet in sight, I lit out for a place of safety which I soon reached, and after waiting another while for Sigel, I went on to Springfield. I was sorry to leave Sigel behind, in the first place, but I supposed all the time he was close to me until I reached the creek, and then it would have done no good for my company to have remained and been cut to pieces also, as were Sigel and his men, who were ambuscaded and all broken up, and Sigel himself narrowly escaped."

CHAPTER III-THE BATTLE OF WILSON'S CREEK.—Concluded

THE SOUTHERN SIDE OF THE STORY.

As one side, the Northern, or Federal, or Union side, of the battle of Wilson's Creek has been told it is but proper that the other, the Southern or Confederate, or secession side, should be given. The statements herein made have been derived from the most authentic sources possible to be consulted. The writer returns his sincere thanks to those Confederate officers, scattered from the Iowa line to the Rio Grande, who have responded to his request for information so promptly and so fully, and in such well written letters.

THE PART TAKEN BY M'CULLOCH'S ARMY.

It will be remembered that Gen. McCulloch had at last yielded to Gen. Price's persistent and positive demands, and had agreed to march against Lyon at Springfield on the night of August 9th and attack him on the morning of the 10th. The march was to be made in four columns and to be begun at 9 o'clock at night.

Just after dark a light rain fell, and it was very dark and a heavy rain storm seemed to be coming up. McCulloch well knew that many of the Missouri troops were not supplied with cartridge boxes, or cartridges either, and that if they moved out from under shelter and it rained hard, as it promised to do, their ammunition would become wet and unserviceable, carried, as much of it would be, in powder-flasks, cotton sacks and shot-pouches. There was also danger that in the Egyptian darkness that had settled down over the land the marching columns would get lost or bewildered, and not come up to the proper place at the proper time. Accordingly, just as some of the troops were preparing to start, McCulloch countermanded the order to march at that time, and the army lay down to sleep, holding itself in readiness to move, however, the men with their guns by their sides. Not much sleep was had, however, for lack of all proper accommodations, and because of the myriads of mosquitoes on the warpath that night up and down the valley of Wilson's creek.

Had Gen. Price been left to himself the day of the 9th, he would have taken "my Missouri boys" that night and marched toward Springfield over the very route that Lyon took from Springfield to the Confederate camp, *via* the Mt. Vernon road and over the prairie, and the two armies, Price's and Lyon's, would have met, to each other's surprise, about midnight, somewhere near the present site of Dorchester.

In his official report to the Confederate Secretary of War, Gen. McCulloch states that his effective force at the battle of Wilson's Creek was 5,300 infantry, 6,000 cavalry, and fifteen pieces of artillery. The majority of the cavalry were armed only with rifles, revolvers, shot-guns, and old flint-lock muskets. There were hundreds of other horsemen along with the army, that were so imperfectly armed as to be of but little efficiency, and during the battle were only in the way.

THE FEDERAL ATTACK.

Col. T. L. Snead states that on the night of the 9th he sat up all night at Gen. Price's headquarters, which were on the side of the creek, at the foot of the sloping, rocky, black-jack hills on whose summit the main battle was fought. About daybreak Gen. Price got up in great impatience and sent for McCulloch, who soon afterward arrived, accompanied by Col. James McIntosh (of the 2d Arkansas Mounted Riflemen), his assistant adjutant-general. "Gen. Price and I were just sitting down to breakfast," says Col. Snead, "and they sat down with us."

As the officers were eating, a messenger came running up from the front, where Gen. Rains' division was posted, a mile or more away, and said that the Yankees were advancing, full 20,000 strong, and were on Rains' line already, peppering his camp with musketry. "O, pshaw," said McCulloch, laughingly, "that's another of Rains' scares," alluding to the Dug Springs affair. "Tell Gen. Rains I will come to the front myself directly," he added. The three officers went on eating, and in a minute or two another messenger came up and reported that the Federals were not more than a mile away, and had come suddenly upon Rains' men as they lay on their arms and had driven them back. McCulloch again said, "O, nonsense! That's not true;" but just then Rains' men could be seen falling back in confusion. Gen. Price rose up and said to Col. Snead, "Have my horse saddled, and order the troops under arms at once." He had hardly spoken when Totten's battery unlimbered and sent its first shot, and about the same instant Sigel's guns opened.

Dispositions for battle were quickly made. Price was ordered to move at once towards Rains with the rest of the Missourians. Pearce was ordered to form on Price's left. Very soon Totten's battery was in plain sight on the top of the hills in front and pounding away, while Sigel's guns in the rear plainly gave notice that the Federals were on all sides.

The surprise was perfect. Most of the Southern troops were asleep. The few pickets that were out had mostly been called in to prepare for the early march, and this enabled Lyon to get close to the line,—upon the skirmishers, in fact,—before being discovered. The troops hurried out as fast and as best they could. The majority of Price's Missourians had their horses with them. Nearly every secessionist, upon enlisting, wanted to ride and did ride. The idea of walking was distasteful in more ways than one,—it was laborious, to begin with, and it was considered somewhat plebeian and disgraceful. And the horsemen, so many of them, proved a serious disadvantage to the Southern cause. They stripped the country in many parts of this State and west of the Mississippi, not only of provisions but of forage and provender, cumbered the roads, and often in battle did more harm than good. At Wilson's Creek the horses became frightened and unmanageable, and at one time they and some of their riders came near stampeding the entire Southern army. Hundreds of them tried to escape from the field by the Fayetteville road, but found it held by Sigel and his Germans.

THE FIGHT AGAINST LYON.

The Missourians under Rains were first attacked by Lyon. Rains had his division under arms and in line with commendable promptness. A great many of his men scattered, it is true, but the majority were soon in ranks and fighting the enemy. Rains' division was a large one, including all the men from the populous secession counties of Saline, Lafayette, Jackson, Johnson, and Pettis, and it held that part of the line in front of Totten's battery. Gen. Price instantly ordered the other division commanders,—Slack, McBride, Clark and Parsons,—to move their infantry and artillery rapidly forward to the support of Rains. Rains' second brigade was in the extreme advance, and consisted of some 1,200 or 1,500 men, mounted and dismounted, temporarily under the command of Col. Cawthorn.

Slack's division of Northwest Missourians was the first to come up, and under the personal direction of Gen. Price himself, who had come to the front, took position on Rains' left, and became instantly engaged. In a few minutes afterwards came John B. Clark's division and formed to the left of Slack. Then came M. M. Parsons' division, with Col. Kelly's regiment or brigade at the head, and went into line to the left of Clark. Then came the division of Gen. J. H. McBride, who took position on the left of Col. Kelly and commanded a flank movement on the right of the enemy, which movement was unsuccessful. (It cannot be learned in what part of the field the forces of Gen. A. E. Steen, of the 5th division, Missouri State Guard, did duty.)

In this position, by Gen. Price's orders, and led by him in person at the first, the entire line advanced in the direction of the enemy, under a continuous fire from Lyon's infantry and Totten's battery, until it reached a position within range of its own guns when the Federal fire was returned, the double-barreled shotguns getting in their work now very effectively. After a few minutes steady firing the Missourians were driven back.

M'CULLOCH COMES TO THE RESCUE.

Meantime Gen. McCulloch had hurried to the lower end of the valley where his division was encamped, and the impetuous Texan chieftain speedily brought out of camp Col. Hebert's Louisiana regiment, and McIntosh's Arkansas mounted riflemen, and hastened to the rescue of the Missourians. This force went to the east side of Wilson's Creek and coming up to the fence enclosing Ray's cornfield, the Arkansas riflemen dismounted, and they and the Louisianians leaped over the fence and charged through the corn upon the Federals (Plummer's battalion) and drove them back upon the main line with loss. This fight in the cornfield was one of the severest of the day, and when it was ended many a corn blade and stalk and tassel had been torn with bullets, and many a dead man lay in the furrows. For no sooner had the Federal infantry been driven back than Dubois' battery opened on the Confederates in the field whose surface had never been disturbed by any thing ruder than Farmer Ray's plow. But now it was soon plowed by shot and shell, and death gathered a full harvest where only the husbandmen had reaped before. The two regiments were driven back with some loss and considerable confusion, but soon reformed and taken charge of by McCulloch in person, who led them to another part of the field.

McCulloch had also ordered up Woodruff's battery, which had engaged Totten and was doing excellent service. During the period of the fight in the cornfield, Price's Missourians were endeavoring to sustain themselves in the center, and were hotly engaged on the sides of the height upon which the enemy was posted. Early in the fight, the 1st Regiment of Arkansas Mounted Rifles, which had been driven out of its camp by Sigel, and had formed a few hundred yards to the north, was brought up by Price's order to the support of Gen. Slack, and formed on his left. Here it fought during the battle, led in person by its commander. Col. T. J. Churchill,[11] who had two horses killed under him. The regiment's loss was 42 killed and 155 wounded. One captain (McAlexander) and three lieutenants were among the killed. The 2d Arkansas Mounted Rifles, Col. B. T. Embry, also fought with the Missourians against Lyon, losing 11 killed and 44 wounded.

Then came the "forward and back" period of fighting described in the Federal account, which lasted for hours. Sometimes the advantage was with one party, sometimes with the other. The firing, both of infantry and artillery, was incessant. Many deeds of gallantry and heroism were performed—enough to immortalize the memory of any one of the perpetrators.

One unfortunate thing, brought about by the battle, was the fact that it produced, or rather made conspicuous, a large crowd of liars who are yet wont to brag and bluster about the various deeds of valor they performed at Wilson's Creek, while the chances are that instead of displaying any remarkable quality of bravery or feat of extraordinary value, they were skulking in the bushes or sitting securely under cover somewhere, not firing a gun or harming an enemy. This is true of both sides. Pity 'tis that any man who wore either the blue or the gray should be a liar, but pity 'tis 'tis true. Deeds worthy of Rome or Sparta—aye, worthy of America, *were* rendered that day of battle on Wilson's creek, but these shameless liars one often meets with did none of them.

From nearly every quarter of Missouri had come the Missourians who this day fought under the flag of the grizzly bears and against the stars and stripes. Slack had men from off the Iowa line; John B. Clark had men from the Northeast (properly belonging to Harris' division, not then south of the Missouri) whose homes were in sight of Hannibal and of the great Mississippi farther to the north. Men fought who, when at home, could stand in their door-yards and look westward over on the prairies of the then territory of Nebraska. Many of McBride's division were from Southeastern Missouri, from the swamps of Pemiscot, from the cypress forests of Dunklin. From the cities—from the warehouses, the counting-rooms and the law offices of St. Louis, St. Joseph and other Missouri towns, had come some men to fight against what they believed to be Federal tyranny and usurpation, and for the honor of old Missouri and the rights of the South. And men fought under Price that day whose feet were on "their native heath," whose homes were in this county, in sight of the battle-ground.

And they all fought well, those in line, whether advancing or retreating, firing or falling back. Not any better than the Federals, perhaps, but fully as well. There were some stragglers on both sides—not all of the cowards were in but one army.

When early in the engagement Gen. Clark sent a mile and a half to the rear for his regiment of cavalry, Col. James P. Major, commanding, that officer was attacked by Sigel at the moment of receiving the order and driven back into the woods with all his force. After reforming and starting toward the front where Lyon was, to join their own division, Major's men were all broken up by large bodies of other horsemen, who, seeking to escape from Totten's grape and Dubois' shells and the Kansas men's musket balls, rode through Major's ranks in all directions, dividing the forces and communicating their own terror to those about them, so that the colonel was left with only one company.

Assisted by Clark's adjutant-general, Col. Casper W. Bell, of Brunswick, Chariton county, and Capt. Joseph Finks, the colonel (Major) succeeded in getting up some 300 men with whom he returned to the rear and assisted in the defeat of Sigel. The remainder of those who could be formed into line (and many of them could when they found that the only road leading out of camp was held by Sigel), were taken charge of by Lieut. Col. Hyde and advanced to the front where Lyon was, but while preparing to charge the Federal left they were driven back by Dubois' battery and some infantry.

At last, after Price's line had advanced half a dozen times and been driven back as often, and after the fight had been going on nearly six hours and victory was not yet certain for either side, McCulloch came back from whipping Sigel and brought with him the Louisianians, Carroll's (Arkansas) and the greater portion of Greer's (Texas) cavalry, Col. Tom P. Dockery's 5th Arkansas infantry, McIntosh's 2d Arkansas rifle regiment, under Lieut. Col. Embry, Gratiot's 3d Arkansas regiment, and McRae's regiment. Reid's battery was also brought up.

THE BEGINNING OF THE END.

The terrible fire of musketry was now kept up along the whole side and top of the hill on which the enemy was posted. Masses of infantry fell back and again rushed forward. The summit of the hill was covered with the dead and wounded. Both sides were fighting with all desperation for the victory. Gens. Price and McCulloch were among their men animating them by their voice, their presence, and their example. Price was slightly wounded but would not leave the field.

To relieve the infantry McCulloch resolved to make a diversion in their favor with the cavalry. Accordingly a portion of Carroll's and Greer's regiments, and a mass of Missourians were formed to go up the valley and fall upon the Federal left, but, as before stated, Dubois' battery and the Federal infantry scattered the horsemen before they could get fairly into line.

VICTORY!

At this critical moment, when the fortunes of the day seemed at the turning point, McCulloch ordered forward his reserves and threw them into the scale. Forward came the rest of Pearce's Arkansas division, Gratiot's and Dockery's regiments, on the run and cheering. Into the thickest of the fight and throwing away their "tooth-picks," as their huge knives were called relied solely on their muskets, and did most effective work in the center of the line. Reid's battery was also ordered forward, and Hebert's Louisianians were again called into action on the left of it. Guibor's battery, of Parson's division, opened with canister on the Federals, and terrible was the din and the slaughter.

Now the battle became general and violent and bloody. Hot as a furnace was the hollow in which the Confederates fought, made so by the blazing August sun overhead. Hot as a Tophet it became, made so by gunpowder, and lead and iron, and sweat and blood. Probably no two opposing forces ever fought with greater desperation, as the Confederate line was advanced on the last charge. But Lyon was killed, Totten's battery moved to the rear, and soon the entire Federal force left the field in possession of the Southerners.

The battle ended suddenly, "as quick as a clap of thunder ceases," one describes it, and for some time after the Federals had retreated it was not certain to the Confederates how the battle had gone. Another attack by the blue-coats was expected and prepared for. Gradually the ground in front where Totten's battery had stood was occupied, and then a line of skirmishers, pushing cautiously to the front, discovered that the victory was theirs. No attempt at pursuit was made, although McCulloch had 6,000 cavalry, whose horses were fresh and rested, and had not sweat a hair that day. That the Federals were not pursued, and in their jaded and exhausted condition cut off from Springfield and captured on the high prairies west of town, seems inexcusable, even to this day, to those posted in the facts.

The Federal officers plainly assert that the reason they were not pursued was because the Confederates were so badly hurt themselves that they could not do so; and further it is claimed that had Lyon lived a Federal victory would have been gained, and Price and McCulloch driven from the field. It is certain (on the authority of Col. Snead), that Price wished McCulloch to pursue, but the latter, for reasons of his own, would not. Then Price resumed command of the Missouri State Guard, and then *he* would not pursue, for reasons of *his* own.

M'CULLOCH'S DESTRUCTION OF SIGEL.

When Sigel came upon the southern end of the Confederate camp the troops he encountered were Churchill's Arkansas regiment, Greer's Texas Rangers, and about 700 mounted Missourians under command of Col. James P. Major and Col. Benjamin Brown, of Ray county, the latter the President of the Missouri State Senate. These troops, taken unawares, were speedily pushed back up the valley across the Fayetteville road. It was at this point of the line,—the Confederate right is faced toward the east,—where McCulloch's Confederates were stationed. When Lyon first opened and alarmed the camp, McCulloch hastened back from Price headquarters, and took up two of his best regiments (Hubert's and McIntosh's), to the assistance of his comrade-commander. The absence of these troops weakened the position of McCulloch very materially, and Sigel had matters his own way for a time. Pearce's division of Arkansas State troops were put in position, somewhat in reserve.

When McCulloch became fully aware that the Federal attack on the south or right was so formidable and so fraught with danger to the entire army, he brought back the Louisiana and Arkansas regiments, and forming them with some of Pearce's division, and Major's and Brown's cavalry, advanced to attack Sigel. The Louisianians and McIntosh's regiment had got the worst of it, in the end, in the fight in Ray's cornfield, but they came up to the work now in brave style. The attack was being made on Sigel's and Salomon's regiments, and the four guns of Schaeffer and Schuetzenbach. There was only scattering firing on the part of the Federals, who mistook the character of the advancing hosts. It was no fault of McCulloch's men, however, that Sigel was deceived. The Louisianians were not to blame that they were mistaken for the Iowa regiment because of their dress.[12]

On they came, regardless of the short-sightedness of their foes, and not knowing or caring anything about their enemies' mistakes until they were within almost grappling distance of Sigel's cannon, when they sprang forward, and with one well contrived and well managed charge swept everything before them. Then followed the events heretofore described—the vain attempts to rally—the disorderly panic-stricken flight—the captures and the pursuit. It must not be forgotten that just before the charge was made, Reid's Arkansas battery opened on the unsuspicious Federal Germans, and they were already in confusion when the Confederate infantry and cavalry were precipitated upon them. Capt. Hiram Bledsoe's Missouri battery, from Lafayette county, with "Old Sacramento," a noted 12-pounder, and three other guns, also did effective work against Sigel, under direction of Col. Rosser, or Weightman's brigade.

As soon as Sigel's destruction had been fairly accomplished (which occupied but a few minutes) McCulloch left the flying fragments to be looked after by sundry detachments of the cavalry, and returned with his infantry and a great deal of the cavalry to the assistance of Gen. Price. In the last efforts against Lyon's column, McCulloch's troops took a conspicuous part, as before detailed; and of course but for the part taken by McCulloch's and Pearce's men the victory could not have been won.

AFTER THE FAMOUS VICTORY.

Dies iræ! O, the moaning and wailing that were all over the land west of the great Fathers of Waters when the full tidings of the battle of Wilson's Creek were learned! From Dubuque and Baton Rouge, from Iowa and Texas, from Louisiana and Kansas, and from every county of Missouri, there went up a sobbing prayer from many a household for strength to bear the bereavement of a father, a husband, a brother or a son slain that 10th of August, 1861, down by the beautiful little stream in the Ozarks.

There they lay, strewn all about over the ground, with faces white and waxen, or clotted with blood, these men who had died to please the politicians. In cosy, shady nooks where fairies might delight to dwell; out in the glare of the blazing sun, festering and corrupting; in cornfields with blade and tassel waving above them, in dells and glens, and vales, and on the hillsides—dead men everywhere. With a tiny bullet hole a baby's finger might stop, marring no feature and mangling no limb; with bowels torn out, with faces shattered, heads torn to pieces, handsome countenances distorted into ghastly, grinning objects—dead men everywhere.

Wounded men everywhere. Crawling about, delirious with pain and agony; lying prone and almost motionless, staring up into the blue sky, dying slowly and making no sign; shrieking, groaning, cursing, praying, imploring help, begging for a bandage, for water, lying quietly, laughing even,—wounded men everywhere. In hospitals, under trees, in tents, in houses, in stables, with surgeons probing and cutting and carving and sawing and clumsily bandaging; in ambulances jolting off towards Springfield; limping along to hide and escape another hurt—wounded men everywhere.

Blood everywhere. On the blades and the silks of the corn; on the leaves of the pretty green bushes.

Great drops on the bunch-grass, but not of the dew;

Staining the velvet moss on the hillsides; purpling in puddles in the pathways and by the roadsides; reddening the lucid waters of bonnie Wilson's creek; flecking the wheels of the guns and daubing the stocks of the muskets; clinging in loathsome gouts to the stems of wild flowers— blood everywhere—human blood—and the best blood of the Republic, too.

COMPARATIVE STRENGTH AND
LOSSES OF THE TWO ARMIES.

The strength of both of the contending armies at the battle of Wilson's Creek is here given as nearly as it has been possible to obtain it. It is believed that the Federal strength has been very definitely learned; that of the combined Southern forces has been approximated in regard to two or three commands in McCulloch's division.

FEDERAL STRENGTH.

According to the reports of the company commanders on the morning of the 9th of August, there were in the column that marched under Gen. Lyon exactly 3,721 men of all arms, infantry, cavalry, and artillery, not including the two companies of home guards under Capts. Wright and Switzler.

Sigel's column consisted of 17 companies of infantry (8 of the 3d Missouri and 9 of the 5th Missouri), numbering 912 men; six pieces of artillery, 85 men; and two companies of cavalry, 121 men;—Total of Sigel's column, 1,118.

Total Federal strength, 4,839—with Wright's and Switzler's home guards, 5,000.

CONFEDERATE STRENGTH.

Without giving exact details, Gen. McCulloch says, in his official report to Gen. Cooper, Adjutant General of the Confederate States: "My own effective force was 5,300 infantry, Woodruff's and Reed's batteries, and 6,000 horsemen." Total, about 11,550.

Gen. Price's division was composed of the following subdivisions:—

Division.	Infantry.	Cavalry.	Total.
Gen. J. S. Rains'	1,306	1,200	2,506
Gen. W. Y. Slack's	659	234	884
Gen. J. H. McBride's	— —	— —	605
Gen. M. M. Parsons'	256	406	662
Gen. John B. Clark's (sr.)	376	250	626
	3,193	2,090	5,283
And Bledsoe's and Guibor's batteries, probably	150		
Grand total of Price's Missourians.[13]	5,433		

July 30, at Cassville, Gen. McCulloch reported his force and that of Gen. Pearce, as numbering in aggregate 5,700, "nearly all well armed." (Rebellion Records, vol. 3, series I, p. 622). Gen. Pearce loaned the Missourians six hundred stand of arms. Afterwards, McCulloch received Greer's South

Kansas Texas cavalry of 1,100 men, and one or two independent companies from Arkansas, making his and Pearce's forces combined, number about 7,000 men. In round numbers the Southern troops numbered about 12,000 at the battle of Wilson's Creek; the Federal or Union forces, 5,000.

THE FEDERAL LOSS.

As officially reported, and on file at this day, was as follows:

Command.	*Killed.*	*Wounded.*	*Missing.*
First Kansas Volunteers,	77	187	20
Second Kansas Volunteers,	5	59	6
First Missouri Volunteers,	76	208	11
First Iowa Volunteers,	13	138	4
Capt. Plummer's Battalion,	19	52	9
Company D, 1st Cavalry, Capt. Elliott,	0	1	3
Capt. Steele's Battalion,	15	44	2
Capt. Carr's Company,	0	0	4
Capt. Wood's Company Kansas Rangers,	0	1	0
Capt. Wright's Dade County Home Guard,	0	2	0
Capt. Totten's Battery,	4	7	0
Capt. Dubois' Battery,	0	2	1
Col. Sigel's Regiment, 3d Missouri,	13	15	27
Col. Salomon's Regiment, 5th Missouri,	13	38	15
Total	235	754	102

Of the wounded forty-eight are known to have died of their injuries afterward, making the actual loss in killed 283.

The principal Federal officers killed were Gen. Lyon; Capt. Carey Gratz, 1st Missouri; Capt. A. L. Mason, 1st Iowa.

Wounded.—Gen. Sweeney; Col. Deitzler, 1st Kansas, (twice); Col. Mitchell, 2d Kansas; Lieut. Col. Merritt, 1st Iowa; Lieut. Col. Andrews, 1st Missouri; Adjt. Waldron, 1st Iowa; Capt. Plummer, of the regulars.

CONFEDERATE LOSS—PRICE'S ARMY.

Gen. Slack's Division.—Col. John T. Hughes' brigade, killed 36; wounded 70 (many mortally); missing 30. Among the killed were C. H. Bennett, adjutant of Hughes' regiment; Capt. Chas. Blackwell, of Carroll county, and Lieut. Hughes. Col. Rive's brigade lost 4 killed, and 8 wounded; among the killed were Lieut. Col. Austin, of Livingston county, a member of the Legislature, and Capt. Engart.

Gen. Clark's Division.—Infantry loss, 17 killed and 71 wounded; cavalry loss, 6 killed and 5 wounded. Among the killed were Capts. Farris and Halleck and Lieut. Haskins. Among the wounded were Gen. Clark himself, and Col. Burbridge, both severely, and Capt. D. H. McIntyre, now attorney general of the State.

Gen. Parsons' Division.—Infantry loss, 9 killed and 38 wounded; cavalry loss, 3 killed and 2 wounded; artillery, Guibor's battery, 3 killed and 7 wounded. Among the killed was Capt. Coleman, of Grundy county. Col. Kelly, commanding the infantry, was wounded in the hand.

Gen. McBride's Division.—Total loss, 22 killed, 124 wounded. Among the latter were Col. Foster, and Capts. Nichols, Dougherty, Armstrong, and Mings.

Gen. Rains' Division.—Weightman's brigade, 35 killed 111 wounded. Cawthorn's brigade, 21 killed and 75 wounded. Among the killed were Col. Richard Hanson Weightman, commanding 1st brigade, and Major Chas. Rogers, of St. Louis.

Two other prominent officers were killed,—Col. Ben Brown, of Ray county, commanding cavalry with McCulloch's army, and Col. George W. Allen, of Saline county, of Price's staff. The latter was shot down while bearing an order, and was buried on the field. Col. Horace H. Brand, of Price's staff, was taken prisoner but released soon afterward.

The total of Price's loss, according to the official reports, was—killed, 156; wounded 609; missing 30.

McCulloch's Army.—The losses of McCulloch's army in the aggregate was 109 killed, 300 wounded and 50 prisoners. Among the officers killed were Capt. Hinson, of the Louisiana regiment; Capt. McAlexander, and Adjutant Harper, of Churchill's regiment; Capts. Bell and Brown, and Lieuts. Walton and Weaver, of Pearce's division. Some of the severely wounded were Col. McIntosh (by a grapeshot), Lieut. Col. Neal, Major H. Ward, Captains King, Pearson, Gibbs, Ramsaur and Porter, and Lieuts. Dawson, Chambers, Johnson, King, Raney, Adams, Hardister, McIvor, and Saddler.

The aggregate Southern loss was not far from 265 killed, 900 wounded and 80 prisoners. A little heavier than that of the Federals, owing to the long range muskets and rifles of the latter and their more efficiently served artillery. All agree that the Confederate and secession batteries as a rule were not well handled.

DISPOSITION OF THE DEAD.

The dead at Wilson's Creek were not well disposed of. All were given hasty and rude sepulture. Of course the Confederate slain fared the better, being buried by their own comrades. The Union dead were put under ground as soon as possible, and with but little ceremony. In an old well, near the battle field, fourteen bodies were thrown. In a "sink-hole" thirty-four of their bodies were tumbled. The others were buried in groups here and there, and the burial heaps marked. In many instances, a few Federal soldiers were present when the burials were made, and identified certain graves. Some of the bodies whose graves were so marked, were afterwards disinterred and removed to their former homes. A number of the Federal dead were never buried; this was particularly true regarding Sigel's men. Dr. Melcher says he saw portions of the bodies of the German Federals along the line of Sigel's retreat, several days after the battle, strewn along near the road, having been torn by dogs and hogs and buzzards. Skulls, bones, etc., indicating that at least a dozen corpses had been left above ground were gathered up. The doctor's statement is corroborated by citizens who lived in the neighborhood.

The weather was hot—oppressively so. Putrefaction soon set in; there was a scarcity of coffins and coffin-makers, and coffin-maker's materials, and perhaps the Confederates did the best they could. Their own dead were, in many instances, given imperfect burial.

In 1867, six years afterwards, when the National Cemetery at Springfield was established, the contractor for the removal of the dead bodies of the Union soldiers on the battle ground, took up and removed, and received pay for 183 bodies, as follows: Out of the "sink-hole," 34; out of the old well, 14; from other portions of the field, 135.

THE HOME GUARDS AT SPRINGFIELD.

Back in Springfield there was a large force of Home Guards, numbering about 1,200, under Col. Marcus Boyd, from Greene and adjoining counties, all under arms, and all ready and willing to fight. But Gen. Lyon held their fighting qualities in such poor esteem—having no confidence that any other sort of troops but regulars would fight well—that he had refused to allow them to go to the field, saying that they would break at the first fire and demoralize the rest of the troops, and perhaps cause him to lose the fight.

But in all probability—no reason appearing to the contrary—if these 1,200 men had been taken out to Wilson's Creek they would have fought well—as well as the volunteers, who fought as effectively as the regulars—and perhaps (who knows?) would have turned the scale in favor of the Federals. Gen. Lyon made a mistake, certainly, in not employing against the enemy in his front every man who could be induced to fire a musket; but his anxiety to not leave his rear and base wholly unprotected from a cavalry dash or sudden movement of some sort, led to his leaving this large force in Springfield, which stood in arms all of the forenoon and heard their comrades fighting so hard away to the southwest, and, anxious as they were to go to their relief, were forbidden to do so.

It is related of a certain doughty captain of the Home Guards then and now a resident of Springfield, that on his reporting to Col. Boyd for orders the morning of the battle, the colonel sent him out on the Mount Vernon road, directing him to observe closely the country to the westward and to report promptly every half hour should anything extraordinary occur. In a few minutes after the opening of Totten's battery, back came the captain ambling along on a little brood mare, which he was industriously larruping with a lath, and reining up his steed in front of Col. Boyd, he made a military salute and announced:—

"Colonel Boyd, *Sir*! The *cannings* is a-firing!" As the roar of every gun had been plainly audible to everybody, this was not a very new piece of information, but Boyd replied, "All right, captain; go back to your post."

Flourishing his lath as before, the captain rode away, and promptly in half an hour—still in his hand the lath, which was doing double service, as a sword and a riding-whip—he returned:—

"Colonel Boyd, *Sir*! The cannings is *still* a-firing!" And so every half hour, until the "cannings" had ceased to thunder, when he returned, and making the same military salute, the faithful lath still in his grasp, he announced:—

"Colonel Boyd, *Sir*! The cannings is ceased a-firing!"

THE RETREAT FROM SPRINGFIELD.

Upon reaching Springfield the Federal army rested a brief time and got itself ready for flight. A conference of the principal officers was held, and the command of all the forces given to Col. Sigel, of whom it is reported Maj. Sturgis said he was not altogether successful in attack, but was "h—l on retreat." The citizens were notified, and hundreds of them began packing up and preparing to follow the army. These were Union people who dreaded the approach of the Southern troops. The Home Guards also got ready to move as a part of the army. Many citizens of the county, living outside of Springfield, got their effects together and were ready to go.

A vast amount of money belonging to the bank had been made ready for shipment, by Lyon's order, and was being guarded by a Home Guard company. Merchandise of all kinds was loaded into wagons and certain of the officers "pressed" teams for the occasion to load commissary and quartermasters' stores into.

Sigel's ordnance officer destroyed a considerable quantity of powder because there were no means of transporting it. The 1st Iowa also burned a portion of its baggage for the same reason. The town was full of frightened men, women, and children, wagons, teams, horses, mules, milch cows, soldiers, infantry, cavalry, and artillery, and there was the greatest confusion all of the evening and till long after dark, even up to the time when the hegira commenced. The public square was a perfect jam of cannon carriages, army wagons, farm wagons, buggies, etc.

CARE OF THE UNION WOUNDED.

By 10 o'clock in the forenoon the wounded Federals had begun to arrive from the front, where the battle was raging, with the news that Lyon was driving the enemy at all points, the Union people cheered, and bestirred themselves to take care of the stricken. The new court-house (the present) and the sheriff's residence were taken for hospital purposes, and by midnight contained 100 men; the Bailey house was filled; the Methodist church building was similarly occupied. Ambulances, carriages, butchers' wagons, express wagons, every sort of vehicle with wheels and springs, plied between the battle field and the town all day and until after dark, bringing off the wounded.

Many of the ladies of the town volunteered their services and became hospital nurses. Maj. Sturgis left with Dr. E. C. Franklin, of the 5th Missouri, the sum of $2,500 in gold, with which to purchase supplies for the wounded left behind, to care for Gen. Lyon's body, and for other necessary expenses. This is upon the authority of Dr. Franklin himself. The doctor was given general charge of the Federal wounded.

THE ARMY SETS OUT.

At last all was ready and the army set out for Rolla, with a train of wagons three miles long and a huge column of refugees, men, women, and children, black and white, old and young, in carriages, wagons, carts, on horseback, on foot, "anyway to get away," as it has been expressed. The march was begun at midnight, and by daybreak the head of the column was outside of the county. No attempt was made on the part of the Southern troops to pursue and capture the column with its $2,000,000 in money and stores, and it was not molested in any way—as, it would seem, it should have been. Sigel was not disturbed until near the crossing of the Gasconade.

Before crossing this river Col. Sigel received information that the ford could not be passed well, and that a strong force of the enemy was moving from West Plains towards Waynesville, to cut off the retreat. He was also aware that it would take considerable time to cross the Robidoux and the two Pineys on the old road. To avoid these difficulties, and to give the army an opportunity to rest, Sigel directed the troops from Lebanon to the northern road, passing Right Point, in the southeastern part of Camden county, and Humboldt, Pulaski county, and terminating opposite the mouth of Little Piney, where in case the ford could not be passed, the train could be sent by Vienna and Linn to the mouth of the Gasconade, while the troops could ford the river at the mouth of the Little Piney to reinforce Rolla. To cross over the artillery he ordered a ferryboat from Big Piney Crossing to be hauled down on the Gasconade to the mouth of Little Piney, where it arrived immediately after the army had crossed the ford. Before reaching the ford, however, Sigel had given up the command of the army to Maj. Sturgis, who marched it into Rolla August 19th, where it went into temporary camp, the first encampment being named "Camp Cary Gratz," in honor of the captain of the 1st Missouri, killed at Wilson's Creek. In a few days the Missouri and Kansas troops and the 1st Iowa, whose term of service had long before expired, were sent to St. Louis to be mustered out.

THE CONFEDERATES ENTER SPRINGFIELD.

The battle of Wilson's Creek ended at about noon of August 10; but not until about 11 o'clock of the next day, or nearly 24 hours after the close of the battle, did the first Confederate troops (save a few prisoners), set foot within the town of Springfield. Sturgis, with the remains of Lyon's corps, was not pursued at all. Sigel's "flying Dutchmen" were chased but a few miles, while no attempt at formidable pursuit or to follow up the victory was made by either McCulloch or Price. Whether this was because, as the Federals claimed, that the Southerners themselves were so badly damaged as to be unable to follow the Federals, but had to wait and allow them to go out of the country before moving camp, or whether Gen. McCulloch himself expected to be attacked, or had other good reasons for sitting quietly by, cannot here be stated.

Lyon's body had been sent in. Certain citizens of Springfield had gone from town to the Southern camp, and back and forth had ridden many a man, but no movement was made until late Sunday morning. At about 11 o'clock some Missouri and Texas cavalry rode into town and halted. No pursuit worthy of the name was attempted after the vast crowd of citizens and soldiers and citizen-soldiery making its exodus from Greene county, in some respects like unto that crowd of fugitives led by the Jewish Lawgiver and guided by a pillar of cloud by day and a pillar of fire by night. Soon the town was pretty well filled with troops, and Price and McCulloch came in. The stores were visited and the proprietors interviewed, and there was great activity in mercantile circles for a time; thousands of dollars worth of goods changed hands in a few hours. Everything was paid for on the spot, — *in Confederate or Missouri scrip.*

The 11th was Sunday, but, as Gen. McCulloch remarked, "it was just as good as any other day in war time," and so the troops were distributed around, encampments laid out, and preparations made to permanently occupy the land. On the next day, Monday, the 12th, Gen. McCulloch issued the following proclamation, which was distributed not only through this county but throughout the greater portion of the southern part of the State: —

PROCLAMATION OF GEN. M'CULLOCH.

Headquarters Western Army,
Camp near Springfield, Mo., August 12, 1861.

To the People of Missouri: — Having been called by the Governor of your State to assist in driving the National forces out of the State and in restoring the people to their just rights, I have come among you simply with the view of making war upon our Northern foes, to drive them back and give the oppressed of your State an opportunity of again standing up as free men and uttering their true sentiments. You have been overrun and trampled upon by the mercenary hordes of the North; your beautiful State has been nearly subjugated, but those true sons of Missouri who have continued in arms, together with my forces, came back upon the enemy, and we have gained over them a great and signal victory. Their general-in-chief is slain, and many of their own general officers wounded. Their army is in full flight; and now, if the true men of Missouri will rise up, rally around our standard the State will be redeemed. I do not come among you to make war upon any of your people, whether Union or otherwise; the Union people will all be protected in their rights and property. It is earnestly recommended to them to return to their homes. Prisoners of the Union army, who have been arrested by the army, will be released and allowed to return to their friends.

Missouri must be allowed to choose her own destiny, no oath binding your consciences. I have driven the enemy from among you; the time has now arrived for the people of the State to act. You can no longer procrastinate. Missouri must now take her position, be it North or South.

Ben McCulloch,
Brig. Gen. Commanding.

This proclamation was well received by the people of the county, especially the Union portion, who expected nothing else that they were to be treated with great severity. All looked forward to a season of security, if not absolute peace. It is painful to be compelled to state, however, that Gen. McCulloch's proclamation was not long observed. Despite its declarations Union men were arrested and their property and that of their secession neighbors seized and appropriated whenever it pleased the subordinate Confederate officers to do so.

In connection with his proclamation, and on the same day McCulloch issued the following congratulatory order to the troops under his command over the result of the battle of Wilson's Creek:—

GEN. M'CULLOCH'S ORDER.

Headquarters Western Army,
Near Springfield, Missouri, August 12, 1861.

The General commanding takes great pleasure in announcing to the army under his command, the signal victory it has just gained. Soldiers of Louisiana, of Arkansas, of Missouri, and of Texas, nobly have you sustained yourselves. Shoulder to shoulder you have met the enemy and driven him before you. Your first battle has been glorious and your general is proud of you. The opposing forces, composed mostly of the old regular army of the North, have thrown themselves upon you, confident of victory; but, by great gallantry and determined courage, you have routed them with great slaughter. Several pieces of artillery and many prisoners are now in your hands. The commander-in-chief of the enemy is slain, and many of the general officers wounded. The flag of the Confederacy now floats near Springfield, the stronghold of the enemy. The friends of our cause who have been in prison there are released. While announcing to the army the great victory, the general hopes that the laurels you have gained will not be tarnished by a single outrage. The private property of citizens of either party must be respected. Soldiers who fought as well as you did the day before yesterday cannot rob or plunder. By order of

Ben McCulloch,
General Commanding.

James McIntosh, Capt. C. S. A. and Adjutant General.

General Price was also seized with the proclamation fever and a few days after the occupation of Springfield, that is to say on August 20th, published the following:—

GEN. PRICE'S PROCLAMATION.

To the People of Missouri:—Fellow-citizens: The army under my command has been organized under the laws of the State for the protection of your homes and firesides, and for the maintenance of the rights, dignity and honor of Missouri. It is kept in the field for these purposes alone, and to aid in accomplishing them, our gallant Southern brethren have come into our State. We have just achieved a glorious victory over the foe, and scattered far and wide the well-appointed army which the usurper at Washington has been more than six months gathering for your subjugation and enslavement. This victory frees a large portion of the State from the power of the invaders, and restores it to the protection of its army. It consequently becomes my duty to assure you that it is my firm determination to protect every peaceable citizen in the full enjoyment of all his rights, whatever may have been his sympathies in the present unhappy struggle, if he has not taken an active part in the cruel warfare, which has been waged against the good people of this State, by the ruthless enemies whom we have just defeated. I therefore invite all good citizens to return to their homes and the practice of their ordinary avocations, with the full assurance that they, their families, their homes and their property shall be carefully protected. I, at the same time, warn all evil disposed persons, who may support the usurpations of any one claiming to be provisional or temporary Governor of Missouri, or who shall in any other way give aid or comfort to the enemy, that they will be held as enemies, and treated accordingly.

Sterling Price,
Maj.-Gen. Commanding Mo. State Guard.

August 20, 1861.

It will be observed that the terms of Gen. Price's proclamation differed somewhat from McCulloch's. The latter declared that prisoners of the Union army would be released and allowed to return to their friends, while Gen. Price declared that no man who had taken an active part in the "cruel warfare which had been waged against the good people (*i.e.*, the secession good people) of the State" should be protected in his rights. And yet Gen. Price was as much a friend of the Union people and Union troops as Gen. McCulloch, and showed them as many favors.

JOY AND CONGRATULATIONS.

The news of the battle of Wilson's Creek was received with great joy throughout the Southern Confederacy and everywhere that the Confederate cause had sympathizers, and the event did much for that cause in Missouri, by stimulating recruiting and causing many an undecided individual to come down off the fence and stand on the Southern side. Some time afterward, November 4, 1861, when the "Claib. Jackson Legislature" (as the Legislature that passed the Neosho ordinance of secession was called), was in session at Cassville, it passed the following resolution, introduced by Mr. Goodlett, under a suspension of the rules: —

> Resolved by the Senate, the House of Representatives concurring therein: That the thanks of the State of Missouri are hereby cordially given to Major-General Price and Brigadier-Generals Parsons, Rains, Slack, Clark, McBride, and Steen, and the officers and troops of the Missouri State Guard under their command, and to Brigadier-General McCulloch and officers and the troops of the Confederate States under their command, for their gallant and signal services and the victory obtained by them in the battle of Springfield.

The following resolutions were introduced into the Confederate Congress on the 21st of August, by Mr. Ochiltree, of Texas, and were passed unanimously: —

> Whereas, It has pleased Almighty God to vouchsafe to the arms of the Confederate States another glorious and important victory, in a portion of the country where a reverse

would have been disastrous, by exposing the families of the good people of the State of Missouri, to the unbridled license of the brutal soldiery of an unscrupulous enemy; therefore

Be it Resolved by the Congress of the Confederate States, That the thanks of Congress are cordially tendered to Brig.-Gen. McCulloch and the officers and soldiers of his brave command for their gallant conduct in defeating after a battle of six and a half hours a force of the enemy equal in numbers and greatly superior in all their appointments, thus proving that a right cause nerves the heart and strengthens the arms of the Southern people, fighting as they are for their liberty, their homes and friends, against an unholy despotism.

Resolved, That in the opinion of Congress, Gen. McCulloch and his troops are entitled to and will receive the grateful thanks of all our people.

CHAPTER IV
PROMINENT REGIMENTS AND OTHER SUBORDINATE COMMANDS ENGAGED IN THE BATTLE.

The particular part taken in the battle of Wilson's creek by some of the leading regiments of each side may be of interest, and is here described, the facts being obtained from actual participants—the commanding officers when possible.

CONFEDERATE COMMANDS.
THIRD LOUISIANA INFANTRY.

A considerable portion of the services of this regiment have already been narrated. Aroused by Gen. McCulloch himself, the colonel of the regiment, Louis Hebert, formed the 3d Louisiana and followed the road to Springfield, a short distance to a narrow by-road running north and leading to Ray's corn-field, then held by Plummer's regulars. In front of the corn-field was a dense thicket, through which the regiment advanced, and here it instantly became engaged. At the first fire Sergeant Major Renwick, of the regiment, and Private Placide Bossier, of the "Pelican Rangers No. 1," of Natchitoches, were killed.

The 3d Louisiana jumped the fence, charged, and soon drove Plummer's men from the corn-field. Still advancing, they reached an oat field, and here Dubois' battery opened on them, as did some of the Federal infantry, and the regiment was driven back in some confusion. Col. Hebert ordered it to fall back to the woods higher to the right, but the regiment became separated, and the greater portion—the right wing and some of the left— were formed outside of the field by Lieut. Col. S. M. Hyams, and, by orders of Gen. McCulloch, went down the creek valley to attack Sigel. Col. Hebert succeeded in forming two companies into a detachment of about 100 men and marched in an opposite direction, toward the force under Gen. Lyon. Col. Hebert advanced within about 500 yards of Totten's battery, where he remained in front of the Federal line for nearly half an hour under a severe fire, when the detachment was forced to retire. Again it formed and then marched down and joined the right wing under Lt.-Col. Hyams, which had just returned from defeating Sigel. The entire regiment then moved against the Federal position on Bloody Hill.

The companies led by Lieut. Col. Hyams against Sigel were the "Pelican Rifles," Capt. John P. Vigilini; the "Eberville Grays," Lieut. Verbois; the "Morehouse Guards," Capt. Hinson; the "Pelican Rangers No. 1," Capt. Breazeale; the "Pelican Rangers No. 2," Capt. Blair; the "Winn Rifles," Capt. Pierson; the "Morehouse Fencibles," Capt. Harris; the "Shreveport Rangers," Capt. Gilmore; a few of the "Monticello Rifles," under Sergeant Walcott, and a detachment of Missourians, 75 in number, commanded by a Capt. Johnson. The regiment was conducted across the ford of Wilson's creek and down the valley in front of Sigel's position by Col. James McIntosh.

Arriving in front of Sigel's battery, the regiment formed, and by order of Lieut. Col. Hyams advanced up the steep hill to the charge. Near the brow of the hill Lieut. Lacey, of the "Shreveport Rangers," sprang on a log, waved his sword, and called out to his company, "Come on, Caddo!" Shreveport is in the parish of Caddo, Louisiana. The whole command rushed forward, carried the position, captured the guns, and drove the already panic-stricken Federal Germans in terror from the field. The captured cannon were rolled down the hill, and one piece, with its horses, was taken to the main command.

The "Pelican Rifles" and the "Iberville Grays" were under command of Capt. Vigilini, of the former company. When within thirty or forty yards of Sigel's battery, Gen. McCulloch being in person with the 3d Louisiana, a Federal soldier appeared in plain view on the hill. Gen. McCulloch himself called out, "What troops are those?" The man replied, "Sigel's regiment," at the same time raising his rifle to shoot the general; but Corporal Henry Gentles, of Vigilini's company, had his Mississippi rifle in position and shot the Federal dead in an instant, thus saving Gen. McCulloch's life.

In the assault on Sigel Capt. Hinson and his brother-in-law, Private Whetstone, both of the "Morehouse Guards," were killed, it is said, by the same shot. Among the members of the regiment who distinguished themselves were Color-bearer Felix Chaler, Corporal Hicock (killed), Drum-Major Patterson, Orderly Sergeant Alphonse Prudhomme, Private I. P. Hyams, Corporal Gentles, and Sergt. W. H. Tunnard. The regiment was especially complimented by Gen. McCulloch in his official report.

The 3d Louisiana lost in killed one commissioned officer, one non-commissioned officer and seven privates; total killed nine; wounded, three commissioned officers, six non-commissioned officers and thirty-nine privates; total wounded, forty-eight; missing, three privates. Total casualties, sixty.

FIRST ARKANSAS MOUNTED RIFLES.

The 1st Arkansas Mounted rifles was at the time of the battle of Wilson's creek in the Confederate service, and was commanded by Col. T. J. Churchill. It belonged to McCulloch's division and was encamped at the lower end of the Confederate position. At about breakfast time Sigel's battery and his infantry opened on the regiment, which lay in an open field. Being exposed to a raking fire from cannon and musketry, the regiment fell back into the woods on the north, and there formed under direction of Col. Churchill himself, who, as soon as his alignment had been made, moved down on the wire road in the direction of Springfield.

Having reached the little valley of Wilson's creek, Col. Churchill was met by an aid de camp of Gen. Price, asking for a reinforcement to go to the assistance of Gen. Slack, then being hard pressed. Col. Churchill immediately moved his regiment rapidly forward, under a heavy fire, took position on Gen. Slack's left, and ordered his men to commence firing. Here the 1st Arkansas fought for about four hours, being in front of Totten's battery and the 1st Iowa the greater portion of the time. At times it would advance, then fall back, but its conduct was most admirable, there being but little if any straggling or disorder. It continued to fight until the Federals retreated from the field.

During the engagement Col. Churchill had two horses shot under him. The lieutenant colonel, Matlock, and the major, Harper, of the regiment evinced great bravery and gallantry. Adjutant James Harper, Capt. M. E. Alexander, and Lieuts. Dawson, Chambers, and Johnson were killed; Capts. Ramsaur and Porter, and Lieuts. King, Raney, Adams, Hardester and McIvor were severely wounded, and Capts. Pearson and Gibbs and Lieuts. Saddler, Wair and Head were slightly wounded. The 1st Arkansas suffered more than any other regiment of Southern troops engaged in the battle. Its loss was forty-two killed and one hundred and fifty-five wounded.

In acknowledgment of the conspicuous services rendered by the 1st Arkansas to his army, Gen. Price, a few days after the battle caused the following letter to be sent to Col. Churchill:—

Headquarters Missouri State Guard,
Springfield, Mo., Aug. 15, 1861.

Colonel:—I am directed by Maj.-Gen. Price to thank you, in
the name of this army and of the State of Missouri, for the
very important services which you and your fine regiment

of Mounted Riflemen have rendered during the campaign in this State, and do particularly acknowledge, in the most grateful manner, the eager bravery with which your men met the enemy on the 10th inst.—the constancy with which they fought, and the spirit with which they rushed upon and drove back his disciplined soldiers. Your own gallantry and skill were so conspicuous on that memorable day that every Missourian will always cherish the remembrance of you with pride and gratitude.

I have the honor to be, Colonel,

Your obedient servant,
Thos. L. Snead,
Acting Adjt. Gen.

Col. Thos. J. Churchill, 1st Regt. Ark. Mounted Rifles.

SECOND ARKANSAS MOUNTED RIFLES.

This regiment was led by its commander, Col. James McIntosh, into the battle, and took part with the 3d Louisiana in the fight with Plummer's regulars, in Ray's cornfield, early in the morning. When first attacked it was at breakfast, but, instead of retreating in confusion, rallied at the call of the bugle, mounted, and was marched by Lieut. Col. Benj. T. Embry to the timber on the east side of Wilson's Creek, north of Woodruff's battery. Here it dismounted and stripped for the fight, and Col. McIntosh then appeared and took it into the engagement.

After the fight in Ray's cornfield, Col. McIntosh was sent for by Gen. McCulloch, and Lieut. Col. Embry took command of the 2d Arkansas. The regiment moved across the creek to the west and became engaged with Lyon's men on Bloody Hill. At one time it repulsed a desperate charge, losing heavily in so doing, however. After this, under Col. Embry, the regiment fell back to the creek and rested a short time, preparing to receive a cavalry charge, which it was expected would be made. It then moved up the hill again, but did not become engaged, and soon after the Federals retreated.

The loss of the 2d Arkansas was, in killed, one non-commissioned officer and 9 privates; total, 10. Wounded, one captain, two second lieutenants, eight non-commissioned officers, and 33 privates; total 44. Total casualties, 54.

M'RAE'S ARKANSAS BATTALION.

Upon the opening of the battle in good earnest, or about 6 A. M., this battalion, led by its commander, Lieut. Col. D. H. McRae, moved up the wire road toward Springfield, and formed to the left of the 3d Louisiana, and in front of Woodruff's battery. In a short time, by Gen. McCulloch's orders, it countermarched and moved off across the valley toward the southwest to take and hold an eminence in that quarter threatened by Sigel's men. While on its way the battalion was broken up by a large body of mounted Missourians, who, panic-stricken and demoralized, were riding rapidly away from Totten's battery and the Federals on Bloody Hill. These mounted warriors rode wildly through the battalion, threatening to trample down the men and forcing them to scatter to save themselves. Col. McRae was able to take but one entire company and a few files of another into the fight proper.

On arriving at the summit of the hill the battalion was fired on by a battery reported as being Bledsoe's, of the Missouri troops, but in reality was Sigel's. Thick brush intervening, Col. McRae was unable to distinguish for himself, but at last charged at a "trail arms." Within twenty paces of the Fayetteville road a body of men were observed moving rapidly away, and these were fired on. Here Col. McRae halted and formed his men so as to sweep the road. In a short time another body came up, and being dressed like the Confederates, and some of them calling out, "We are from the South," deceived Col. McRae until nearly all of them had passed, when he opened fire on their rear. He then led his men to the hill where Sigel's battery had been captured, and here he found the three companies of the battalion that had been cut off by the Missouri horsemen. The united battalion then marched to the Fayetteville road to the north, it having been reported that the Federals were reforming there, but this report was found to be untrue, and Col. McRae returned to camp.

The loss of the battalion was two men killed, one mortally wounded, one severely wounded, and five slightly wounded.

THIRD ARKANSAS INFANTRY.

This regiment was commanded by Col. John R. Gratiot, and belonged to the Arkansas State troops, Gen. Pearce's division. Its lieutenant-colonel, David Provence, and its major, Ward, were present at the battle.

In the first part of the action the regiment was moved to the support of Woodruff's battery, and here it remained for some hours under a heavy fire of shot and shell. At about 11 o'clock Gen. Pearce ordered the regiment to cross the creek and move to the help of Price's division. Col. Gratiot marched the men over the stream and up the ridge by a flank movement and in column of fours. When near the Federal position the line was fronted and faced the enemy, and moved forward, but just then a heavy fire was opened in front, two guns of Totten's battery turned loose on the regiment with grape, and canister and shell, and so terrific was the ordeal, that the regiment was obliged to lie down and return the fire in that position. This was the last fight of Lyon's men, and they kept it up only about thirty minutes, when they retreated. The 3d Arkansas remained on the field in position long after the firing had ceased.

After Maj. Sturgis had retreated with the remnant of Lyon's division from Bloody Hill, it was feared that he would cross the creek, move round to the east and come upon Woodruff's battery, still in position on the ground it had occupied during the day. Col. Gratiot's regiment was again ordered to the support of this battery, and here it remained until ordered into camp by Gen. McCulloch.

The regiment suffered most during the thirty minutes it was engaged with the Federal infantry (Second Kansas and part of First Iowa), and Totten's battery on Bloody Hill, but it stood well and gave back blow for blow. Capts. Brown and Bell were killed and about twenty-five other brave men and true met their fate in this battle.

Capt. Woodruff's battery, the "Pulaski artillery," was attached to the Third Arkansas during the battle. This battery did more execution and service than any other Confederate battery that took part in the engagement. The damage it inflicted on the enemy was prodigious. Officers and men behaved with great coolness, courage and judgment.

The casualties in the Third Arkansas and Woodruff's battery were:— killed, twenty-five; wounded, eighty-four; missing, one; total, one hundred and ten.

FOURTH ARKANSAS INFANTRY.

The Fourth Arkansas infantry on the morning of the battle was placed under Adjutant-General Rector, who remained in command during the day. The regiment was not brought into immediate action, being stationed on the hill for the protection of Reid's battery, and, although exposed to the trial of having to submit to a severe fire from the enemy, which it was unable to return, all the officers and men behaved with great coolness during the day. There were none killed or wounded in this regiment. [The colonel of the First Arkansas, J. D. Walker, is at present (1883) one of the United States Senators from Arkansas.]

FIFTH ARKANSAS INFANTRY.

This, another regiment of Gen. Pearce's division, was commanded by Col. Tom P. Dockery, and for about two hours after the battle commenced was posted on the height southeast of McCulloch's encampment, and occupied a hill east of Wilson's creek as a guard for Reid's battery.

When the Third Louisiana and Third Arkansas moved up against the Federals on Bloody Hill Col. Dockery sent to their support Capts. Titsworth's, Dismuke's, Neal's, Dowd's, Whaling's and Lawrence's companies, all under Lieut. Col. Neal. While gallantly leading his men Col. Neal fell severely wounded, and Col. Dockery then assumed command. Only the companies named were actively engaged against the enemy on Bloody Hill, the companies of Capts. Hartzig, Arnold, McKeon and Hutchinson having been detailed to serve as skirmishers at one time after Reid's battery had changed position. The Fifth Arkansas did its duty well in this battle, and its conduct was commented on by Gen. Pearce in the warmest terms. It never wavered or showed the least sign of demoralization. The loss of the regiment was three killed and eleven wounded.

FIRST ARKANSAS CAVALRY.

Mention has already been made of the services performed by this regiment. For a while it supported the Missourians of Price's division; then it charged by the flank on Totten's battery; then charged again on the position held by the Second Kansas, and all the time during the engagement was under fire. While thousands of other cavalry were demoralized and fleeing hither and thither, the First Arkansas kept on the field and sought more than once to charge as cavalry over ground almost Alpine in character—rugged, rough, precipitous and broken. Its commander, Col. DeRosey Carroll, was complimented more than once for the gallant conduct shown by himself and his regiment.

The loss of the First Arkansas cavalry was five killed; two mortally wounded; twenty-six severely wounded, and nineteen missing, as follows:—

Capt. Lewis' company—two killed, two mortally and five severely wounded.

Capt. Park's company—one killed, three wounded, one missing.

Capt. Walker's company—four wounded, including Capt. Walker himself, and three missing.

Capt. Withers' company—two killed, four wounded, two missing.

Capt. Perkins' company—four wounded, four missing.

Capt. Kelly's company—one missing.

Capt. Armstrong's company—one wounded, eight missing.

FEDERAL COMMANDS.

THE FIRST IOWA INFANTRY.

The 1st Iowa Infantry was a three months regiment whose time had expired several days before the battle, but it had remained on duty with Lyon to aid him in his emergency. At the time of the fight its colonel, J. F. Bates, lay sick in Springfield, and Lieut. Col. Wm. H. Merritt, led the regiment. As has been stated the 1st Iowa was in Lyon's column. At the beginning of the fight it was in the reserve, but advanced when the 1st Kansas gave way and received the first fire of the enemy while in a state of some confusion, the result of the retreat of the Kansans through the Iowa's ranks. The regiment fired on a body of Confederate cavalry advancing to charge Totten's battery, and dispersed it, or drove it back. Soon after the regiment became engaged generally, and bore its full share of the conflict.

Four companies of the 1st Iowa, and a company of regular infantry under Capt. Lothrop, supported Totten's battery at the close of the engagement and covered the retreat, receiving and returning the last fire of the enemy. The regiment lost 13 killed (including Capt. A. L. Mason) 138 wounded and 4 missing. Total 155.

FIRST MISSOURI INFANTRY.

The 1st Missouri Infantry was led into the battle of Wilson's Creek by its lieutenant colonel, George L. Andrews, its colonel, Frank P. Blair, being in his seat in Congress at the time. When Gen. Lyon's column had reached the immediate proximity of Rains' division, the regiment was brought forward to the head of the column and directed to march parallel with the advance—Gilbert's regulars—and about 60 yards distant to the right. In a few moments orders were received to throw one company forward as skirmishers, and Company H, Capt. Yates, was sent forward, followed by the regiment in column of companies.

It is claimed that the action was begun by shots from Capt. Yates' skirmishers. At any rate, soon after they opened fire Company B was sent up as a reinforcement, and the regiment wheeled into line and immediately became engaged, at first returning a fire directed against its left flank. Very soon after Woodruff's and Guibor's batteries opened on the Federal position and their shells fell uncomfortably plenty among the 1st Missouri. One or two of the shells which did not explode were examined and pronounced to be those furnished Sigel's batteries, leading to the conclusion that Sigel was firing by mistake against Lyon's column.

The regiment stood well in line and fought bravely. Capt. Nelson Cole was severely wounded in the jaw, but remained on the field, and, though unable to speak, from the nature of his wound, he continued to encourage his men by signs to stand their ground. Capt. Cary Gratz, of St. Louis, a native of Lexington, Ky., while advancing at the head of his company, discovered a body of Confederates advancing, led on by a mounted officer. Capt. Gratz fired with his revolver and the Confederate officer fell from his horse, but rose and rushed toward his lines, when the captain fired again and the other officer pitched headlong to the ground. Almost immediately Capt. Gratz fell dead, being pierced by four or five shots.

Capt. John S. Cavender, with his company, G, was in an advanced position and several times prevented the left flank from being turned. Col. Andrews, while with the left wing was severely wounded, but he procured a big drink of whiskey, and soon returned to his post. In a few minutes his horse was killed and fell upon him. Going to Dubois' battery during a lull in the fighting, Col. Andrews was sent to the rear by Surgeon P. M. Carnyn.

The Confederates now pressed the 1st Missouri so vigorously that the regiment in all probability would have given way had not Maj. Schofield and Gen. Lyon opportunely brought up the 1st Iowa and Maj. Osterhaus (assisted by Lieut. David Murphy, of the 1st Missouri) come forward with his battalion of the 2d Missouri Infantry. The regiment then remained on the field and did nobly during the remainder of the engagement, and when ordered to fall back with the main column and leave the field it did so in good order.

Out of 27 officers who went into the fight, 13 were either killed or wounded. Capt. Madison Miller discovered a movement of the Confederate cavalry to his rear and stopped it by the stout fight he made with his company, assisted by the artillery. Capt. Cavender, though severely wounded, refused to leave his post, mounted his horse and remained until completely exhausted. Surgeon Carnyn, on more than one occasion, took up a musket and fought in the ranks. Lieut. David Murphy, although severely wounded in the leg, went to the rear and assisted in bringing up Osterhaus' battalion of the 2d Missouri. Adjt. Hiscock and other officers bore themselves so well that they received especial mention.

Among the men Corporal Kane, of Company K, when the color sergeant was killed and nearly all the color guard either killed or wounded, brought the colors safely off the field. Sergt. Chas. M. Callahan, Company K, Sergt. Chris. Conrad, of Company G, and Private Elworthy, of Company F, were noted for their valuable services and for their coolness and bravery. The part borne by the 1st Missouri may be imagined when it is remembered that its loss was 76 men killed, 208 wounded and 11 missing.

FIRST KANSAS INFANTRY.

On coming upon the battle field in the early morning, the 1st Kansas Infantry—led by its colonel, Geo. W. Deitzler, and its major, John A. Halderman—was posted in the rear of the 1st Missouri and 1st Iowa. Very soon Gen. Rains' skirmishers or outposts were driven in, Totten's battery took position and opened fire, while the 1st Missouri was sent up and soon became engaged.

At this time, under an order from Gen. Lyon, the 1st Kansas moved to the front in "double quick," while the right wing and one company from the left, under command respectively of Capts. Chenoweth, Walker, Swift, Zesch, McFarland, and Lieut. McGonigle—all under Col. Deitzler— advanced to a position beyond that occupied by the 1st Missouri, and here, forming in the very face of the enemy, engaged the Confederates and held their ground steadfastly under an uninterrupted and murderous fire of artillery and infantry.

The four remaining companies of Capts. Clayton, Roberts, Stockton, and Lieut. Angell, under Maj. Halderman, having been posted on the right of Totten's battery as support, where they suffered severely from a constant fire, were now brought up by Maj. Halderman, who called out, "Forward boys, for Kansas and the old flag!" Aligning with remarkable coolness upon the remnant of the six right companies the four left companies settled down to work. With but slight and immaterial changes of position the 1st Kansas occupied this ground for over two hours, holding its ground and dealing and receiving severe punishment.

While thus engaged, Capt. Chenoweth, Capt. Clayton and a portion of Capt. McFarland's company under Lieut. Malone, were ordered to charge the enemy with their commands, which order they executed and drove back the Confederates a considerable distance, although soon after they were themselves compelled to retire. While leading this charge Col. Deitzler had his horse killed under him, and was himself severely wounded. The command then devolved on Maj. Halderman. The regiment now had a very exposed position, lying in plain view, obliquely across a ridge, but, though it suffered severely, it bore itself well.

When the 2d Kansas fell back the 1st formed on its left, three companies remaining on the brow of the hill, and on the right of the battery. After the severe volley firing had ceased for a few minutes—the Confederates having retired—it was recommenced by them again as they advanced, and kept up for nearly a quarter of an hour, when the whole Federal line, apparently, opened on them and they again retired down the hill. After this Maj. Sturgis ordered the retreat.

With scarcely any material change of its position the 1st Kansas stood under fire and returned it, maintained every ground assigned it, without turning its back on the foe, for the five long hours during which the battle raged. Its loss was the heaviest in killed of the Federal regiments engaged—77, one more than the 1st Missouri. It had 187 men wounded and 20 missing; total, 284. It went into action with nearly 800 men, and left the field in good order with about 500.

THE SECOND KANSAS INFANTRY.

This regiment, as stated in the general description of the battle, was at first stationed in reserve on a hill on the right of and overlooking Ray's cornfield, where Plummer's battalion fought. After Plummer had been driven back and the pursuing Confederates checked by Dubois' battery, Lieut-Col. Chas. W. Blair, of the 2d Kansas, rode to Gen. Lyon and requested that the regiment be given a place in the front. Gen. Lyon gave the order and Col. Mitchell brought the regiment forward, in time to take part in the last grand charge. Prior to this, and early in the action, before the regiment, as a regiment, was fairly under fire, a force of Missouri cavalry (presumably of Rains' division) attempted a flank movement, and Maj. W. F. Cloud, of the 2d, taking Capt. McClure's company and deploying it, drove them back after a few volleys.

As the regiment went up to the forefront, Gen. Lyon put himself at its head and assisted the field officers in bringing it into action. Just as the regiment raised the crest of the hill, and while it was still in column, a terrific fire was opened on it, and it was under this fire that Gen. Lyon fell dead and Col. Mitchell was severely wounded. Gen. Lyon was leading the 2d when he was killed. After Col. Mitchell was wounded, command of the regiment was assumed by Lieut-Col. Blair and Maj. Cloud, who threw the men into line, and after a hard fight of fifteen minutes the Confederates were driven down the hill, and a lull in the conflict resulted.

About this time Capt. Powell Clayton's company, of the 1st Kansas, was attached to the left of the 2d, and the companies of Capts. Roberts, Walker and Zesch, also of the 1st, were formed on the right. On the right of this position a ravine stretched down to the enemy, and up this ravine the Confederates (of John B. Clark's division) attempted to flank Col. Blair. Some men sent down it from Capt. Cracklin's company did not return, and then Col. Blair himself rode out to see what was the matter. He had not gone twenty yards when he "found what darkened de hole!" A sharp fire was opened on him and his horse killed, but the colonel himself was unhurt, and mounting a horse brought him by his orderly, Alex. H. Lamb, he was soon again directing the movements of his men.

Meantime, aware of the danger in front, Maj. Cloud had gone back to Sturgis for reinforcements and obtained two guns of Totten's battery, under Lieut. Sokalski. These came up in good time. As they stopped, Capt. Chenoweth, of the 1st Kansas, rode out to the head of the ravine, and saw the Confederates coming up it in considerable numbers. Cloud and Sokalski got the guns in position and opened on the ravine. As the Confederates approached nearer Col. Blair ordered the men to lie down and load and fire in that position and to be careful of their ammunition. Here the men received a most terrific fire, which they seemed to relish. Artillery and musketry were playing on them, but the shot and shell went too high and only the grape, the muskets, and the rifles of the enemy did execution. Yet not a man broke ranks or left his place in the line. At last the Confederates fell back or slackened their fire and the artillery limbered up and retired to the rear to join in the general retreat, which bad been ordered some minutes before.

Maj. Sturgis, on assuming command after Lyon had fallen, sent Col. Blair word to retreat as soon as he could do so with safety, and after the Confederates had fallen back the last time he did so. The men were brought off in good order and in slow time, without a panic or confusion. After crossing the first ravine in the rear the line was reformed and marched by the right flank to the main command and off the battle field.

The loss of the 2d Kansas in the battle was 5 killed, 59 wounded, and 6 missing—total, 70. Both officers and men behaved splendidly. When Col. Mitchell fell he turned over the command to Col. Blair, saying: "Colonel, take the regiment and maintain the honor of Kansas." As he was being carried from the field he called out to Gordon Granger, of Sturgis' staff, "For God's sake, support my regiment." Of Lieut. Col. Blair, it was said by Gen. Sturgis that "he attracted the attention of all who saw him." Col. Mitchell, Lieut. Col. Blair and Maj. Cloud were all highly complimented by Gen. Sturgis and Gen. Fremont and recommended for promotion. Maj. Cloud, Adjutant Lines and Capt. Ayres were mentioned in Col. Blair's report as conspicuous for their gallantry.

TOTTEN'S BATTERY.

The share of fighting done by this organization (Light Company F, 2d Artillery), in the battle of Wilson's Creek was large and important. Soon after the skirmishers of Lyon's advance fired on the Southern pickets, the line of march, as directed by Gen. Lyon in person, lay through a small valley which debouched into that through which Wilson's Creek runs at the point immediately occupied by the front of Price's troops and just where a road to Springfield entered the valley, keeping along the foot of the hills, and soon after the battery opened. The left section, under Lieut. Sokalski, was first brought to bear upon the enemy in the woods in front, and shortly afterward the other four pieces were thrown forward into battery to the right on higher ground. A few rounds from the artillery assisted the infantry in driving the secession troops back toward the crests of the hills, nearer and immediately over their own camp.

Capt. Totten now conducted the battery up the hill to the left and front, and soon found a position where he brought it into battery directly over the northern position of the enemy's camp. The camp of Gen. Rains' division lay directly beneath the front and to the left of, though very close to, the position of the battery, while a battery of the secessionists (Woodruff's Arkansas) was in front and within easy range. Of course Rains' camp was entirely deserted, and therefore Totten's first efforts were directed against the Arkansas battery in his front and right. The right half of Totten's guns were principally directed against Woodruff, although the entire six pieces as opportunity offered, played upon him. The two batteries pounded away on each other for some time, neither seeming to get much the advantage of the other. As the position of the Arkansas battery was somewhat masked by the timber, Totten's gunners were obliged to give direction to their pieces by the flash and smoke of the opposing artillery.

In the meantime, while this fight between Totten and Woodruff was in progress, the battle was raging in the thick woods and underbrush to the front and right of the position of the Federal battery, and the 1st Missouri was being hard pressed.

Gen. Lyon ordered Totten to move a section of his battery forward to the support of the Missourians, which was done, the guns coming up on a run and unlimbering in front of the right company of the regiment. A Confederate regiment with a Confederate flag, which at that distance seemed to be the stars and stripes, was two hundred yards away, and fearing they might be friends Totten hesitated before opening. Their fire soon undeceived him and he turned loose his guns upon them with canister from both pieces.

The next important step in the progress of the battle was when a portion of Clark's (?) division tried to force its way up the neighborhood road passing along toward Springfield in order to turn the Federal right. For a time the situation was critical, for the Missourians were plucky and were fighting hard. Four pieces of Totten's battery were still in position commanding that point, and Dubois' four guns were on the left also near the road and commanding it. As Slack's men came in good view and range, both artillery and infantry opened on them and drove them back.

Just after this had been done Gen. Lyon came up to the battery and complimented the men who were working it. Capt. Totten saw blood trickling from Lyon's heel and the general said he had been wounded in the leg, but not seriously. The captain offered his commander some brandy in a canteen, but the general refused it, and rode away, and that was the last time Totten ever saw Lyon alive. Soon after leaving Totten, Lyon sent him word to support the Kansas men on the extreme right, who were being hard pressed. Lieut. Sokalski took up his section immediately and saved the Kansans from being overthrown and driven back.

After this came an attempted charge on the Federal position by some Missouri, Texas and Arkansas cavalry, of which there was a great abundance. Some 800 of them, including a battalion or so of Greer's Texans and Carroll's Arkansans, fresh from the southern end of the little valley where Sigel had been so easily whipped, formed at the foot of the hill on which four of Totten's guns stood and were getting ready for a charge, when the artillery and the infantry opened on them and they were driven away so rapidly that they were out of sight in a moment.

The last point where the battery was engaged was on the right of the left wing of the 1st Iowa and somewhat in front. Lyon was then killed and Sturgis was preparing to retreat. Totten's battery was still in action, two pieces in advance on the right so hot that the water thrown on them almost hissed, and yet pounding away. The left wing of the 1st Iowa came up and supported the guns from the field, and they were brought away off the field and to Springfield, without the loss of a sponge-stick. The battery lost 4 killed and 7 wounded; no prisoners.

DUBOIS' BATTERY.

This battery so named consisted of four pieces of light artillery, three six-pounders, and one twelve-pounder, commanded by second Lieutenant John V. Dubois, of the U. S. Mounted Rifles, a semi-cavalry regiment. Lieut. Dubois had been detailed from his regiment to command this battery, newly organized and manned by recruits.

Entering the fight Dubois selected his own position, some 80 yards to the left and rear of Totten's battery, where his men were partially and his horses entirely protected from the enemy's fire. He assisted Totten in clearing the ground of the enemy at the start, and under direction of Capt. Gordon Granger (afterward a major general), opened on McCulloch's men down in Ray's cornfield, who had just driven back Plummer, and drove them in disorder, Capt. Granger directing one of the guns. The Confederates rallied behind a house (J. A. Ray's), on the right of their line. Dubois struck this house twice with a twelve-pound shot, when a hospital flag was displayed and he ceased firing. Using small charges of powder, Dubois' guns now shelled the thicket in the ravine, a short distance in front, and forced some of Price's Missourians to retire.

Bledsoe's battery now opened on Dubois from the crest of the hill opposite and "Old Hi's" fire did great execution. "Old Sacramento," as Bledsoe's twelve-pounder was called, got in her work very disadvantageously to the Federals. The shots passing over Dubois' gun fell among the Federal wounded that had been carried to the rear and did considerable execution. Dubois could not entirely silence Capt. Bledsoe's guns, but he made it very uncomfortable for him. One shot from Dubois' gun, killed two of Bledsoe's battery horses, tore off one arm and all of the hand of the other arm of the man who held them, Judge James Young, now of Lexington, and killed another man far in the rear. While engaged with Bledsoe and his Lafayette county battery, Lieut. Dubois assisted in driving back the cavalry that formed to charge on Totten.

During the last charge of Price and McCulloch on Sturgis, two of Dubois' guns were limbered up to be sent to Totten, but before a road could be opened through the brush and through the wounded, orders came for Dubois to fall back to a hill in the rear and protect a retreat. This he did, remaining until all the troops had passed when he turned and marched toward the rear. Shortly after starting back the twelve-pounder broke down. While it was being repaired Maj. Osterhaus's two companies remained with it to protect it, and then followed in its rear until the main portion of the command on the prairie was reached. Here the battery joined Steele's battalion and formed the rear guard the rest of the way into Springfield, neither firing or receiving a shot on the way and not being molested in anywise—never seeing an enemy. Being well protected during the entire engagement, the loss in this battery was very slight—none killed and only two severely wounded. Several of the men and Dubois himself received slight wounds.

STEELE'S BATTALION.

The battalion of regulars commanded by Capt. Fred Steele was composed of two companies of the 2d regular infantry—company B, commanded by Sergt. Griffin, and company C, commanded by Sergt. McLaughlin, a company of "general service" recruits under Lieut. M. L. Lothrop, and a company of mounted rifle recruits commanded by Lance-Sergeant Morine. It will be seen that Capt. Steele had but one *commissioned* officer under him.

During the early part of the action the battalion was in position to support Dubois' battery, but had no opportunity of engaging the enemy except to assist in dispersing a body of cavalry that threatened the rear. Soon after the fall of Gen. Lyon, Capt. Gilbert's company joined the battalion and Maj. Sturgis ordered Capt. Steele to form in line of battle and advance against the enemy's front. Heavy firing on both sides followed, without any apparent permanent advantage to either, until the suspension of hostilities mentioned before. During this suspension Lieut. Lothrop took his company forward as skirmishers, but they were driven back in very short order and without much ceremony.

A Confederate field piece (probably one of Guibor's), was run up under the hill and threw grape and occasionally a shell over Steele's battalion, but with no serious effect, as the shots passed too high. Two other pieces were added and worked vigorously but not carefully, and with no other effect than to cause Steele's men to lie close to the ground.

In the last grand charge on Totten's battery and the main Federal position, Steele's battalion did good work, the men firing away nearly all their cartridges. Just before the retreat began Capt. Gordon, with his hastily collected detachment from different regiments, and Capt. Steele repulsed another attack, and enabled Totten's battery and other commands to retire in good order. On the retreat to Springfield after reaching the prairie Capt. Steele commanded the rear guard, and states that he was not molested at all, "never seeing an enemy." The loss of Steele's battalion was 15 killed, 44 wounded, and two prisoners. Sergt. Morine, commanding the rifle recruits, was killed on the field.

PLUMMER'S BATTALION.

This battalion performed brave service at the battle of Wilson's Creek. It belonged to the 1st U. S. Regular Infantry, and most of the men had been some time in the service. Frequent reference has already been made to the part it performed in this battle. Gilbert's company had the advance upon reaching the battlefield and was the first thrown forward on the skirmish line. The principal portion of the battalion, commanded by Capt. J. N. Plummer himself, made the fight in Ray's cornfield against the 3d Louisiana and 2d Arkansas Mounted Rifles, and afterward engaged in the terrible conflict on and along Bloody Hill.

The battalion was remarked as much for its coolness as for its bravery. Upon the retreat of the Federals to Springfield it entered the town in perfect order, the flag flying, the drums beating, and the men keeping perfect step as if they were on parade or drill, and as collected and unexcited as if nothing of consequence had taken place that day.

The battalion was composed of Cos. B, C and D, of the 1st Regulars, commanded by, respectively, Capts. Gilbert, Plummer, and Huston, and Lieut. Wood's squad of Rifle Recruits. Capts. Plummer and Gilbert were severely wounded, and Capt. D. Huston then took command. Out of 230 men engaged, the battalion lost 19 killed, 52 wounded and 9 missing—a total of 80, or a little more than one-third of the entire number in the fight.

THE HOME GUARDS.

Two companies of mounted Union home guards—one called the Dade county squadron, commanded by Capt. Clark W. Wright, and the other under Capt. T. A. Switzler—were present at the battle, but took no very important part. They made some charges on scattered squads of secessionists, driving them under cover and out of all danger to the Federal line, but for the greater portion of the time they were stationed to the right and rear of Lyon's position as a post of observation and to prevent the line from being flanked by the enemy's cavalry. In one of the charges Capts. Wright and Switzler ascended a hill in plain view of Gen. Rains' camp, and counted a number of tents.

DISPOSITION OF THE BODY OF GEN. LYON.

Ah, Sir Launcelot. Thou there liest that never wert matched of earthly hands. Thou wert the fairest person and the goodliest of any that rode in the press of knights. Thou wert the truest to thy sworn brother of any that buckled on the spur; * * * and thou wert the sternest knight to thy mortal foe that ever laid spear in rest.

For the purpose of ascertaining the truth concerning the death and burial of the body of Gen. Lyon, the writer hereof caused certain newspaper publications to be made in the St. Louis *Republican* and other journals, making inquiries pertinent to the case. Many and varied were the replies, some of which, perhaps, ought to be given, as illustrating the different lights in which men see the same object, and the morbid desire for notoriety on the part of others, which leads them to lie like book-agents, in order that their names may be published in connection with some notable event. No less than ten newspaper articles were published and thirty-two written communications were received by the compiler relating to the death and burial of Gen. Lyon. The result was forty-two different versions thereof.

The work of ascertaining the truth was thereby complicated instead of being facilitated. A dozen or more claimants for the distinction of having first discovered the body on the battlefield appeared. Half a score bore the corpse to Gen. Price's tent. Twenty saw the body, noted its appearance carefully, etc. Knowing from incontrovertible proof how the general was dressed when he was killed, the writer inserted a test question asking that his garb be described. Two ex-officers, one Union, the other Confederate, answered that he was "in full general's uniform." A minister of the gospel, who was also the "*first* to discover the body," promptly replied that he was "dressed in a complete suit of black broadcloth, white shirt, gold studs, fine boots and kid gloves!" The majority of the answers, however, were to the same effect, that he was dressed in his old fatigue uniform of his former rank—that of captain in the regular army—without epaulettes or shoulder-straps. After much labored investigation the writer has ascertained the following facts, which he can easily substantiate:—

Gen. Lyon was killed while placing the 2d Kansas Infantry in position, by a rifle or navy revolver ball through the region of the heart. He was borne to the rear by Lieut. Schreyer, of Capt. Tholen's company, 2d Kansas, two other members of the same regiment, and Ed. Lehman, of Co. B, 1st U. S. Cavalry, the latter the soldier who caught the general's body as it fell from the horse. As the body was borne to the rear, Lieut. Wm. Wherry, one of the general's aides, had the face covered, and ordered Lehman, who was crying like a child, to "stop his noise," and tried in other ways to suppress the news that the general had been killed. The body was placed in the shade of a small black-jack, the face covered with half of a soldier's blanket, the limbs composed, and in a few minutes there were present Surgeon F. M. Cornyn, Maj. Sturgis, Maj. Schofield, Gen. Sweeney, and Gordon Granger, and perhaps other officers. Cornyn examined the body, and from the side of the face wiped the blood made by the wound in the head. He also opened the vest and split the general's shirts, which were soaked with blood, and examined the wound, which was found to be in the heart, the aorta having been pierced. The minister's story, which he has the effrontery to give a newspaper publication, relates that two or three hours afterward the body was neatly dressed, with its smooth *white* shirt and studs, kid gloves, etc.! Those best acquainted with the personal habits of Gen. Lyon say he never wore a pair of *kid* gloves during his term of service.

Maj. Sturgis ordered the body to be carried back to a place selected as a sort of field hospital and there to be placed in an ambulance and taken to

Springfield. While the body was here lying a few Federal officers examined it and one of them reports that the face had again become bloody, from the wound in the head, and that the shirt front was gory from the death wound. About twenty minutes after the body had been brought back, Lieut. David Murphy, of the 1st Missouri, who was already badly wounded in the leg, and Lehman placed the body in an army wagon, being used as an ambulance, and belonging to Co. B, 1st U. S. Cavalry. This wagon was about to start to Springfield, and contained already some wounded men. A few minutes later a sergeant of the regular army came up and ordered the body taken out, saying, "There will be an ambulance here in a minute for it." The corpse was then carried beneath the shade tree where it had before reposed.

The Federal army now retreated, and the ambulance ordered never came up. Before the Confederates came on to the ground where the body lay, which location was 200 yards northeast of "Bloody Hill," half a dozen slightly wounded Federal soldiers had gathered about the dead hero, and an hour after the Federal retreat a party of Arkansas skirmishers came upon them and discovering the occasion of the crowd instantly spread the news that Gen. Lyon had been killed. Immediately there was a great tumult and the report was borne to Price and McCulloch by half a dozen. Many were incredulous and did not believe that a body so plainly dressed—in an old, faded captain's uniform, with but three U. S. buttons on the coat and a blue (or red) cord down the legs of the trousers to indicate that he was in the military service—was that of Gen. Lyon.

The body was then placed in a small covered wagon, used as an ambulance, to be conveyed to Gen. McCulloch's headquarters (not Gen. Price's) when an order arrived that it should be taken to Price's and delivered to Dr. S. H. Melcher, of the 5th Missouri, who, as before stated, had come upon the field in company with Dr. Smith, Gen. Rains' division surgeon. Dr. Melcher had been informed by Col. Emmett McDonald that Lyon had been killed, and at once asked for his body. When the little covered wagon containing the corpse had driven up and Gen. Price and Gen. Rains and other officers had viewed the body, it was turned over to Dr. Melcher. A number of Southern soldiers standing by drew knives and made attempts to cut off some buttons or pieces of the uniform as relics, and one or two expressed a wish to "cut his d——d heart out;" but Gen. Rains drew his sword (or revolver) and swore he would kill the first man that touched the corpse, and Emmett McDonald denounced the ruffianly would-be violators in the harshest terms—and McDonald could be harsh when he wanted to be!

Beside the body of Gen. Lyon was a wounded man, who was now taken out, and then Gen. Rains himself and some of his cavalry escorted the wagon to the house of Mr. Ray, on or near the battlefield. It is proper now to give the testimony of Dr. Melcher himself, as given to the writer and furnished the press for publication. Speaking of the courtesy of Gen. Rains in escorting the body to Ray's house, Dr. Melcher goes on to say:

> Arriving there the body was carried into the house and placed on a bed; then I carefully washed his face and hands, which were much discolored by dust and blood, and examined for wounds. There was a wound on the right side of the head, another in the right leg below the knee, and another, which caused his death, was by a small rifle ball, which entered about the fourth rib on the left side, passing entirely through the body, making its exit from the right side, evidently passing through both lungs and heart. From the character of this wound it is my opinion that Gen. Lyon was holding the bridle rein in his left hand, and had turned in the saddle to give a command, or words of encouragement, thus exposing his left side to the fire of the enemy.

> At this time he had on a dark blue, single breasted captain's coat, with the buttons used by the regular army of the United States. It was the same uniform coat I had frequently seen him wear in the arsenal at St. Louis, and was considerably worn and faded. He had no shoulder-straps; his pants were dark blue; the wide-brim felt hat he had worn during the campaign was not with him. After arranging the body as well as circumstances permitted, it was carried to the wagon and covered with a spread or sheet furnished me by Mrs. Ray.

> When I was ready to start Gen. Rains said: "I will not order any to go with you, but volunteers may go;" and five Confederate soldiers offered their service of escort. One drove the team; the others, being mounted, rode with me in rear of wagon. The only name I can give is that of Orderly

Sergt. Brackett of a company in Churchill's Arkansas regiment. Another of the escort was a German who in 1863 was clerking in Springfield, and during the defence of Springfield against the attack of Marmaduke, January 8, 1863, did service in the citizens' company of 42 men which was attached to my "Quinine Brigade" from the hospitals.

The following is a copy of a paper written at Mr. Ray's house. The original I now have:—[14]

Gen. James S. Rains, commanding Missouri State Guards, having learned that Gen. Lyon, commanding United States forces during action near Springfield, Mo., Aug. 10, 1861, had fallen, kindly afforded military escort and transportation subject to my order. I have also his assurance that all the wounded shall be well taken care of and may be removed under the hospital flag, and that the dead shall be buried as rapidly as possible.

[Signed]

S. H. Melcher,

Asst. Surg. 5th Reg. Mo. Vols.

Wilson Creek, Aug. 10, 1861.

The above fully approved and indorsed.

[Signed]

James S. Rains,

Brig.-Gen. 8th M. D., M. S. G.

About half way to Springfield I saw a party under flag of truce going toward the battlefield. Arriving at Springfield, the first officer I reported to was the ever faithful Col. Nelson Cole, then captain of company E, 1st Missouri Volunteer Infantry, who, with what remained of his gallant company, was guarding the outposts. I passed on to the camps of Gen. James Totten and T. W. Sweeney. Here Gen. Totten relieved my escort and sent them back to their command, a new driver was furnished, and I delivered the body of Gen. Lyon to Maj. J. M. Schofield, 1st Missouri Volunteer Infantry—now Maj.-Gen. Schofield, U. S. A.—at the house that had been used previous to the battle by Gen. Lyon for his headquarters.

It is proper to state that Dr. Melcher's testimony is corroborated in part by two survivors of the 1st Arkansas, and by Mrs. Livonia Green, now of Lane county, Oregon, and also by Mrs. Jerome Yarbrough, of this county, both of the latter being daughters of the Mr. and Mrs. Ray mentioned. (Mr. and Mrs. Ray are now dead.)

After Sturgis' army had gotten well on the road to Springfield, it was discovered that Gen. Lyon's body had been left behind. Sturgis immediately started back a flag of truce party under Lieut. Canfield, of the regular army, with orders to go to Gens. Price and McCulloch, and, if possible, procure the remains and bring them on to Springfield. Lieut. Canfield and party went to the battle field, saw Gen. McCulloch, obtained his order for the body (the general remarking that he wished he had a thousand other dead Yankee bodies to send off) and there ascertained that the body had already started for the Federal lines.

When the corpse was deposited in the former headquarters of the general, on the north side of College street, west of Main, in Springfield, word was sent to Sturgis. He and Schofield and other officers held a consultation, and decided that the body should be taken with the army to Rolla, if possible. There not being a metallic coffin in the place, it was determined to embalm it, or preserve it by some artificial process. Accordingly, the chief surgeon, Dr. E. C. Franklin, was sent for. Responding to the inquiries of the writer, Dr. Franklin says:—

> About ten o'clock p. m., on the night when it arrived at head quarters, I was summoned there and then first saw the body of Gen. Lyon lying upon a table, covered with a white spread, in a room adjoining the one where two or three of the Union officers were seated. Gens. Schofield, Sturgis, and others consulted me as to the possibility of injecting the body with such materials that would prevent decay during its transit to St. Louis. I prepared the fluid for injection into the body, but discovered that instead of being retained within the vessels it passed out into the cavity of the chest. This led me to suspect a laceration either of one of the large arteries near the heart, or, possibly, a wound of the heart itself. This hypothesis, coupled with the fact that there was an external wound in the region of the heart, confirmed my opinion of the utter uselessness of attempting the preservation of the body during its passage to St. Louis.

These facts I reported to the commanding officer, who then gave me verbal orders to attend to the disposal of the body in the best possible manner. At this time preparations were being made and the orders given for the troops to retreat and fall back upon Rolla, some fifty or more miles nearer St. Louis. Returning to the general hospital, of which I was in charge, I detailed a squad of nurses to watch by the body of Gen. Lyon till morning, which order was faithfully carried out. I then disposed of my time for the best interests of the wounded and sick under my charge.

Dr. Franklin was furnished with money and directed to have the general's remains well cared for, and he ordered an undertaker, Mr. Presley Beal, to make a good, substantial coffin at once. Early the following morning, in some way, word was sent to Mrs. Mary Phelps, wife of Hon. John S. Phelps, that the body of the great Union leader was lying stiff, and bloody, and neglected in the temporary charnel house on College street. Soon she and the wife of Mr. Beal were by his side, and watching him. Not long thereafter came the wife of Col. Marcus Boyd and her two daughters (one of whom, now Mrs. Lula Kennedy, still resides in Springfield), and kept them company. And so it was that women, "last at the cross and first at the tomb," were those who kept vigil over the corse of the dead warrior, who, although he died the earliest, was one of the greatest Union generals the war produced.

The body had now lain about twenty-four hours in very hot weather. It was changing fast, and its condition made it necessary that it should be buried as soon as possible. Mrs. Phelps left Mrs. Boyd and her daughters and went to see about the coffin. Dr. Franklin came in and sprinkled the corpse with bay rum and alcohol. Mr. Beal brought the coffin, and soon a wagon—a butcher's wagon—was on its way to Col. Phelps' farm with all that was mortal of the dead hero, and with no escort save the driver, Mrs. Phelps, Mr. Beal and one or two soldiers.

Col. Emmett McDonald, than whom the war produced no more knightly a soldier, had been made a prisoner by Gen. Lyon, at the capture of Camp Jackson. When Lyon was killed, Col. McDonald not only assisted Dr. Melcher in recovering the body, but Dr. Franklin says of him:—

> Here let me do justice to Col. Emmett McDonald, who called upon me at the general hospital and after some conversation

in regard to the circumstances attending the death of Gen.
Lyon, tendered to me an escort of Confederate troops as
a "guard of honor" to accompany Gen. Lyon's remains to
the place of burial, which I refused from a too sensitive
regard for the painful occasion, and an ignorance of military
regulations touching the subject.

Mrs. Phelps was practically alone at the time. Her husband was in his seat in the Federal Congress, her son, John E. Phelps, had followed off the Federal army, and even her faithful servant, George, had accompanied his young master. But Mrs. Phelps was a lady not easily daunted, or one that would shrink from what she considered a duty, no matter how unpleasant it might be. The body was taken to Mrs. Phelps' residence, and not buried at once, it being the understanding that it would be sent for soon. Mr. James Vaughan, who owned a tin-shop in Springfield, was ordered to make a zinc case for the coffin, to assist in the preservation of its contents.

The coffin was temporarily deposited in an out-door cellar or cave, which in summer had been used as an ice-house, and in the winter as an "apple-hole," and was well covered with straw. It was here placed about two o'clock on the 11th. A day or two later, the slave, George, returned. While the body of Gen. Lyon lay in Mrs. Phelps' cellar, the place was visited by some citizens and many Southern soldiers. It is much to be regretted that some brutes there were among the soldiers that treated the remains of the dead man with all disrespect, cursing them and him openly and in the vilest terms. One young officer is reported to have said to Mrs. Phelps: "There is quite a contrast betwixt the resting place of old Lyon's body and his soul, isn't there, Madame? The one is in an ice-house; the other in hell!" he added with a heartless chuckle.

At last some drunken ruffians, by threatening to open the coffin and "cut the d—d heart" of the body for a relic, so frightened Mrs. Phelps, causing her to fear that the remains would be mutilated in some horrible manner, that she asked Gen. Price to send a detail and bury the body. This was done by volunteers from Guibor's and Kelly's infantry, of Gen. Parsons' division, at that time encamped on Col. Phelps' farm. It is believed the body was not *buried* until the 14th. The slave, George, dug the grave, which was in Mrs. Phelps' garden. Some of the soldiers stamped on the grave in great delight. An Irishman told Capt. Guibor, "Be jabers, we shtomped him good."

On the 22d of August there came to Springfield a party in a four-mule ambulance, bearing with them a 300-pound metallic coffin. This party was

The next day the party left Springfield and were in Rolla on the 25th and in St. Louis the 26th. Here a military escort joined. From thence the party proceeded to Eastford, Connecticut, the birthplace of the general, which place was reached September 4th, there being great receptions and honors paid the body in the cities and towns *en route*. September 5th the body was buried in the family burying ground at Eastford. "Upon the coffin, as it lay in the Congregational church when the funeral ceremonies were being rendered," says Mr. Woodward, who was present, "were placed the hat, a light felt, which the general had waved aloft when rallying his ranks at Wilson's Creek, and also the sword, scarred and weather-beaten from sharing in the long hard service of its owner." The hat was brought from the battlefield by the wounded men in the wagon in which the general's body was first placed, and was given to Mr. Hasler by the driver, who had preserved it. Both hat and sword were given to, and since have been in the possession of the Connecticut Historical Society.

Gen. Lyon was born in Eastford, Connecticut, July 14, 1818. He entered West point in 1837; graduated in 1841, standing eleventh in a class of fifty. He served in Florida in 1841-2; was in the Mexican war under Taylor and Scott; in California and on the frontier from 1850 to 1861. He was never married. The statement that he bequeathed his private fortune to the Federal government is erroneous.

FOOTNOTES

[1]   The fight at Dug Springs was called by some of the Confederate officers, derisively, "Rains' Scare."

[2]   The following is a literal copy of the memorandum given to Col. Phelps by Gen. Lyon, when the former left Springfield. Lyon instructed Phelps to give this to Fremont: *"Memorandum for Col. Phelps.*—See General Fremont about troops and stores for this place. Our men have not been paid and are rather dispirited; they are badly off for clothing and the want of shoes unfits them for marching. Some staff officers are badly needed, and the interests of the government suffer for the want of them. The time of the three months volunteers is nearly out, and on their returning home my command will be reduced too low for effective operations. Troops must at once be forwarded to supply their place. The safety of the State is hazarded. Orders from Gen. Scott strip the entire West of regular forces and increase the chances of sacrificing it. The public press is full of reports that troops from other States are moving toward the northern border of Arkansas for the purpose of invading Missouri.*Springfield, July 27."*

[3]   From statements of two prominent Union men of Greene county who were present.

[4]   Afterward Major General in command of the Federal troops in Arkansas.

[5]   Gen. Sweeney said: "Let us eat the last bit of mule flesh and fire the last cartridge before we think of retreating."

[6]   There are grounds for stating that Lyon knew of the intended attack upon him within four hours after it had been agreed upon, receiving his information through one of his spies, actually a commissioned officer in the Missouri State Guard!

[7]   It must be borne in mind that the Confederate line extended in a general direction from the north to the south along Wilson's Creek; that Lyon attacked the northern end from the west and northwest, while Sigel was stationed at the southern end, over a mile away.

[8]   Afterward Major General and in command of this department.

[9]   Which was done near Mr. Robinson's.

[10]   It was not Totten's battery, but Reid's Confederate battery, from Ft. Smith, Ark. It was well supplied with grape from the Little Rock arsenal.—Compiler.

[11]   Since Governor of Arkansas.

[12]   At the breaking out of the civil war, the color of the infantry uniform of the U. S. army was gray. Upon its adoption by the Confederates this color was changed, and blue substituted.

[13]   Gen. A. E. Steen's division seems to have been attached to McCulloch's army. It was insignificant in numbers.

[14]   The writer has seen and carefully examined the original of this paper. It is written in pencil, but is quite legible. The handwriting of Gen. Rains was identified beyond question. The paper was kindly furnished by Dr. Melcher for the purposes of this history.

www.ingramcontent.com/pod-product-compliance
Lightning Source LLC
LaVergne TN
LVHW041735190726
843493LV00008B/2355